THE POLYVAGAL THEORY

*A Self-Help Guide To Understanding
The Nervous System - Ease Anxiety,
Depression, Stress, PTSD By
Activating The Vagus Nerve*

By

Isaac Anderson

TABLE OF CONTENTS

UNDERSTANDING VAGUS NERVE

The vagus nerve is the body's superhighway, which carries information between the brain and internal organs. During periods of rest and relaxation, it regulates the body's response. The broad nerve emerges from the brain and extends in several ways, to the neck and the chest, where it brings sensory information from the skin of the head, regulates the muscles you use to breathe, talk, and affects the immune system. According to the Encyclopedia Britannica, the vagus is the 10th out of 12 cranial nerves straight from the cortex. Even though we call the vagus nerve singular, it's a pair of nerves that simply originate from the left and right side of the medulla oblongata part of the brain stem. According to Merriam-Webster, the nerve gets its name from the Latin word wandering as the vagus nerve is the biggest and most ramified cranial nerve. Through circling and spreading across the body, the vagus nerve has primary control over the parasympathetic division of the nervous system: the rest-and-digest contraposition to the fight-or-flight response of the sympathetic nervous system. The vagus nerve sends signals that regulate heart and breathing levels and improve digestion when the body is not under threat. During times of stress, regulation shifts to the nervous mechanism, resulting in the reverse reaction. The vagus nerve also takes feedback impulses from the internal bodies back to the brain, enabling the brain to track the behavior of the heart. The vagus nerve travels from the spine to the forehead and from the thorax to the belly. It is a combined nerve containing parasympathetic fibers. There are two sensory ganglia (nerve masses carrying electrical impulse) in the vagus nerve: the upper and lower ganglia. The roots of the upper ganglion encircle the tissue in the ear concha. The lower ganglion contains two branches: the pharyngeal nerve and the upper laryngeal nerve. The recurrent laryngeal nerve extends the lower neck and upper thorax from the vagus to inside the laryngeal muscles (voice box). The vagus also contains the nucleus, esophagus, and lung divisions. The vagus internalizes most of the digestive tract and intestinal viscera in the uterus. The body has 12 cranial nerves. These come in pairs to help connect the brain to other parts of the body like the back, neck, and torso. Others give sensory information to the brain, providing knowledge about smells, colors, tastes, and sounds. Such nerves have motor functions. Certain cranial nerves control the movement and function of certain glands and different muscles. These are also engine features. While some cranial nerves have visual or motor functions, others have both functions. The vagus nerve is a muscle like that. The cranial nerves are categorized into Roman numerals based on their location. The vagus nerve is also referred to as the cranial nerve X. The word "vagus" means to walk in the Latin language. This is a very suitable name because the vagus nerve is the longest cranial nerve. This runs from the brain stem to the colon portion.

A nerve that supplies pharynx (throat), larynx (voice box), trachea (windpipe), coronary, respiratory, esophagus, and intestinal tract nerve up to the transverse part of the intestine. The vagus nerve frequently sends sensory information from the ear, mouth, pharynx, and larynx to the brain. The 10th cranial nerve is the vagus nerve. The medulla oblongata, part of the brain cortex, originates and stretches from the brain trunk to the colon. A complete loss of the vagus nerve causes a common symptom in which the palate declines on the side of the injury, and the gag reflex on the other side of the nerve is also damaged. The accent is rough and nasal, and the vocal cord is immobile on the affected side. The consequence is trouble swallowing and communicating (dysphagia). The vagus nerve has several major branches, including the recurring laryngeal nerve. The digestive system includes large branches of the vagus nerve. Around 10% to 20% of the vaguely connected nerve cells transmit instructions from the brain to regulate muscles that pass the food through the gut. Such organs are then regulated by a different nervous system inside the walls of the digestive system. The other 80% to 90% of the nerves produce sensory information from the stomach and gut to the cortex. The brain-gut axis of this communication line between the brain and gastrointestinal tract keeps the brain aware of the state of muscle contraction, level of movement of food through the body, and sensations of appetite or satiety. A study published in the Internal Medicine Journal in

2017 showed that the vagus nerve is so similar to the digestive system that nerve stimulation can intensify irritable bowel syndrome. Throughout recent decades, several researchers have discovered that this brain-good axis is different— the bacteria that reside inside the intestines. According to a 2014 study published in Advances in Experimental Medicine and Biology, this microbiota contact with the brain via the subtly nerve influences not only the food consumption but also mood and inflammation. Much of the current research includes studies using mice and rats instead of humans. Nonetheless, the findings are compelling and demonstrate that microbiome changes will cause brain changes. The vagus nerve has the highest cranial nerve range. His pharyngeal and laryngeal branches send motor signals into the pharynx and larynx; his heart branches slow the pace of heartbeat; his bronchial branch functions in order to constrict bronchi; and his esophagus regulate unintended muscles in the esophagus, intestine, gallbladder, pancreas and small gut, peristaltic sensations and gastrointestinal secretions. Vagus nerve stimulation, in which the neuron is activated through electric pulses, is often used for epileptic or depressed individuals who are not normally treatable, and treatment for the condition of Alzheimer's disease and migraine has also been investigated.

The vagus has traditionally been identified as an efferent nerve and as a sympathetic antagonist of the nervous system. Many tissues get parasympathetic effects by the splanchnic nervous system and sympathetic effects. The parasympathetic nervous system is responsible for controlling vegetative processes together with the sympathetic nervous system by behaving in relation to each other. The parasympathetic innervation allows the blood vessels and bronchioles to dilate and the salivary glands to intensify. Conversely, sympathetic internation causes blood vessels to become constricted, bronchioles to dilate, the heart rate increases, and the bowel and urinary sphincters are shortened. The stimulation of the parasympathetic nervous system increases bowel motility and glandular secretion in the gastrointestinal tract. In comparison, the sympathetic action decreases bowel activity and blood flow into the gut, which enables a higher blood supply to the heart and muscles when the patient has existential stress. The vagus nerve is the main component of the parasympathetic nervous system and controls a wide variety of essential functions, including mood control, immune response, metabolism, and heart rate. This provides one of the ties between the brain and the gastrointestinal tract and sends information through afferent fibers to the brain about the state of the inner organs. We discuss various functions of the vagus nerve in this review book, which makes it attractive for the treatment of psychiatric and gastrointestinal disorders. Preliminary evidence is provided that the activation of the vagus nerve is a potential alternative therapy for medication-refractory depression, post-traumatic stress disorder, and inflammatory bowel disease. Treatments aimed at the vagus nerve increase the vagus sound, and reduce the production of cytokine. Both are important resistance mechanisms. Stimulation of distinctly afferent fibers in the intestine affects the brain stem monoaminergic structures, which play a key role in severe psychiatric conditions such as depression and anxiety disorders. In particular, preliminary evidence exists for a beneficial mood and anxiety influenced by intestinal bacteria, partially by altering the function of the vagus nerve. Because the vagal tone is associated with the ability to regulate stress response and is activated by breathing, increased tolerance, and mitigated effects of mood and anxiety is likely caused by meditation and yoga.

The vagus nerve enters the medulla Oblongata in the groove between the olive and the lower brain peduncle and exits the skull through the middle of the jugular foramen. The vagus nerve in the neck provides the most pharynx and larynx muscles a necessary innervation, which is responsible for swallowing and vocalizing. The thorax contains the heart's strongest part of sympathy and induces a reduction of the heart rate. The vagus nerve guides smooth muscle relaxation and glandular secretion in the gut. Vagal efferent fiber pregnant neurons arise from the dorsal motor nucleus of the vagus nerve situated in the medulla and converge in the lamina propria and muscular externa in tissue and tendon layers of the intestines. The celiac branch supplies the intestine to the distal portion of the descending colon from proximal duodenum. Mucosal mechanoreceptors, chemoreceptors, and pain receptors in the esophagus, stomachs, and the proximal intestines are included as well as sensory endings in the liver and pancreas. Among ganglia doses, the visual afferent cell bodies transmit data to the nucleus tractus solitarii (NTS). The NTS describes the vagal sensory information in various CNS regions such as locus coeruleus, ventrolateral rostral medulla, amygdala, and thalamus. The vagus nerve regulates internal organ processes, such as appetite, heart rate, respiratory rate as well as vasomotor action, as well as certain reflexes such as coughing, sneezing, swallowing, and vomiting. The activation allows acetylcholine (ACh) to be produced at the synaptic junction of secretive cells, intrinsic nerve fibers, and smooth muscles. ACh binds to muscarinic and nicotine receptors and activates muscles in the parasympathetic nervous system. Animal studies have shown impressive vagus nerve regeneration ability. Induced temporary detachment and regeneration of the main vagal afferents as well as synaptic plasticity of the NTS, for example, subdiaphragmatic vagotomy. Furthermore, 18 weeks after subdiaphragmatic vagotomy, vagal afferents recovery in rats can be accomplished, even when the efferent reinnervation of the gastrointestinal tract is not recovered after 45 weeks.

Vagus Nerve as a Link between the Central and ENS

The relation between the CNS and the ENS, known as the brain-good axis, enables a two-way link between the brain and the gastrointestinal tract. It regulates physiological homeostasis and binds the emotional and cognitive areas of the brain to peripheral intestinal roles, including immune regulation, intestinal permeability, entero-endocrine signals. The head / gut axis comprises the spine, the spinal cord, the adaptive nervous system, and the hypothalamic-pituitary-adrenal axis. The vagal efferent sends "back" from the brain to the intestine by efferent fibers, which account for 10–20% of all fibers and vagal afferents "up" from the intestinal wall to the spine, containing 80–90% of all fibers. The vagal afferent trajectories are used to activate the HPA axis that controls the organism's adaptive responses to stressors of every sort. Environmental stress and higher systemic proinflammatory cytokines stimulate the HPA axis by secreting the CRF from the hypothalamus. The activation of CRF facilitates the secretion of adrenocorticotropic hormone (ACTH) from the hypophysis. In effect, this activation results in the secretion of cortisol from the surreal glands. Cortisol is an essential stress hormone that affects many human bodies, including the brain, bones, muscles, and fat.

The communications channels of both the neurological (vagus) and hormonal (praxis) enable the brain to control the activities of cells with intestinal functional effectors, including immune cells, epithelial cells, enteric neurons, smooth muscle cells, Cajal interstitial cells, and enterochromaffin cells. On the other hand, these cells are affected by intestinal microbiota. The gut microbiota has an important effect on the axis of the brain that not only communicates locally with intestinal and ENS cells but also directly influences the neuroendocrine and the metabolic processes. Emerging data support the role of microbiota in anxiety and depressed behavior. Tests on germ-free animals have shown that microbiota influences stress reactivity and anxiety-like behavior and control the point of reference for HPA action. Such animals also generally show decreased anxiety and increased stress response with elevated ACTH and cortisol levels.

In the case of food consumption of vagal afferents that are inside the gastrointestinal tract, digestible and circulating and processed fuels are easily and discreetly accounted for, while vagal efferents and hormonal pathways codetermine the ingestion, preservation and mobilization rate of the nutrient. Histological and electrophysiological evidence suggests that a number of chemical and mechanosensitive receptors are found in visceral afferent endpoints of the vagus nerve in the intestine. Such receptors are the sources of intestinal hormones and regulatory peptides, which in response to nutrients, distension of the stomach, and synaptic signal releases from the enteroendocrine cells of the gastrointestinal system. These affect food intake management and satiety regulation, gastric emptying, and the energy balance by sending signals originating from the upper intestine to the nucleus of the brain's solitary tract. Most such hormones, including peptide cholecystokinin (CCK), ghrelin, and leptin, are responsive to food contents in the intestines and contribute to controlling short-term appetite and satiety feelings.

Cholecystokinin controls gastrointestinal functions and, by stimulating CCK-1-receptors on vagal afferent fibers that intrude the intestine, prevent gastric emptying and food intake. CCK is also essential for the secretion of pancreatic fluid and the development of gastric acid, gallbladder contraction, gastric emptying, and digestion. CCK release from the small intestine is induced by saturated fat, long-chain fatty acids and amino acids, and small peptides arising from protein digestion. CCK is categorized by the number of amino acids; I .e.CCK-5, CCK-8,CCK-22, and CCK-33. CCK-8 is still the primary type in the neurons, while in endocrine cells, there is always a combination of small and large CCK peptides, often overwhelming CCK-33 or CCK-22. Among rodents, both long-and short-chain fatty acids among food activate but do so by different mechanisms. Short-chain fatty acids like butyric acid have a direct effect on the CCK-dependent terminals. The long-chain fatty acids activate vagal afferents. It seems that exogenous CCK administration prevents endogenous CCK secretion. CCK is also present as a neurotransmitter, in vagal enteric afferent cells, cerebral cortex, thalamus, hypothalamus, base ganglia, and dorsal hindbrain. This stimulates vagal afferent terminals in the NTS directly by increasing the release of calcium. In addition, the evidence is shown that CCK can stimulate neurons in the hindbrain and myenteric intestinal plexus (a plexus that supplies motorcycles to both the muscle layers of the intestine), rats and that vagotomy or capsaicin therapy contributes to a decrease in CCK-induced phos production in the brain (a kind of proto-oncogene). There is also substantial evidence that high CCK levels exacerbate anxiety. Therefore, CCK is used to model anxiety disorders in humans and animals as a threat tool.

Ghrelin is another hormone that is released from the stomach, which plays a key role in promoting the intake of food by inhibiting vagal afferents. The level of ghrelin production is raised by fasting and falling after a meal. Central or peripheral administration of acylated ghrelin to mice effectively increases food intake, and the production of growth hormone and weight gain is triggered by chronic administration. The activity of Ghrelin's on feeding in rats with vagotomy or capsaicin, a particular afferent neurotoxin, is eliminated or alleviated. For mammals, intravenous infusion or injection increases both appetite and food intake as leptin is reduced by ghrelin. It is, therefore, not shocking that obesity and insulin resistance are affected by secretion. In the vagus nerve, leptin receptors were also identified. Rodent studies indicate that leptin and CCK synergistically interact to cause short-term food intake inhibition and long-term body weight loss. The epithelial cells that respond to ghrelin and leptin are situated in the vicinity of vagal mucosal ends and modulate the function of vagal afferents to control food intake together. Leptin is losing its potential effect on vagal mucosal adjuvants after fasting and diet-induced obesity in mice. The gastrointestinal tract is the important interface between food and the human body and, by using a common G-protein-paired taste receptor, can detect specific tastes much the same way as the tongue does. Various flavor qualities cause different gastric peptides to be released. Through inducing CCK activation, bitter taste receptors may be seen as potential targets for decreasing hunger. In fact, activation of bitter taste receptors induces the release of the ghrelin and thus triggers the vagus nerve.

WHAT IS POLYVAGAL THEORY

Porges ' polyvagal hypothesis was established by his vagus nerve experiments. The vagus nerve represents the parasympathetic nervous system, a soothing component of our physiology of the nervous system. The parasympathetic part of the autonomous nervous system parallels the sympathetic active component, but in far more nuanced ways than was commonly thought.

The three-part nervous system

Our nervous system was historically described as a two-part antagonistic system, with a higher number of stimulation signals less relaxation and soothing signals fewer activation. Polyvagal theory describes a different kind of nervous system reaction called Porges, the social engagement system, which is a playful combination of stimulation and relaxation, based on a particular nervous effect.

The system of social interaction allows us to establish relationships. Helping our consumers to use their social engagement mechanisms allows them to make their management strategy more adaptive.

The other two elements of our nervous system help us withstand life risks. The two defense mechanisms caused by these two portions of the nervous system are already well known to most counselors; sympathetic fight-or-flight and parasympathetic shut-down also called freeze-or-faint. On the other hand, the use of our social engagement system requires a sense of security.

Polyvagal theory helps us to understand that both distinct nerve branches relax the muscle but in different ways. Shutdown, or freeze-or-faintness, takes place through the vagus nerve branch. This reflex will sound like the sore muscles and lightness of a weak grip. It can push us to immobility or dissociation when the dorsal vagal nerve shuts down the body. In comparison to the effects on the heart and lungs, the dorsal branch controls the function of the body below the diaphragm.

The ventral branch of the vagal nerve affects the body over the diaphragm. This is the division that represents the system of social responsibility. The ventral vagal nerve dampens the active state of the body—an image of a horse leading it back to the barn. You will continue to pull the reins on and loosen them in complex ways to ensure that the horse holds pace properly. The ventral vagal nerve also permits complex activation and provides a different level of activation than sympathetic activation.

The vital vagal release takes milliseconds, while sympathetic activation takes a few seconds and requires many chemical reactions that are close to removing the kidneys of the horse. Therefore, once the chemical reactions to battle and travel have ended, our bodies will need 10–20 minutes to return to our pre-flight / pre-flight state. These forms of chemical reactions do not require a vagal release into operation. We can thus make faster transitions between stimulation and relaxation, as we can do when using the kidneys to regulate the animal.

If you go to a dog park, you'll see some terrified puppies. We exhibit fight or flight behavior. Some dogs may show that they want to play.

This symbol also takes the form that we human beings have stolen in yoga for the downward-looking posture. If a dog gives this warning, it leads to an extreme level of excitement. This playful drive, however, has a distinct nature than the strength of fight or flight behavior. This amusing nature characterizes the system of social participation. We work from our social engagement framework as we view our world as secure.

Three organizing principles are at the heart of the Polyvagal Theory.

Hierarchy: The autonomic nervous system reacts through three channels to stimuli in the body and feedback from the environment. Such pathways function in a certain order and respond in predictable ways to challenges. In evolutionary order, the three pathways (and their receptive patterns), from the oldest to the newest, consist of the dorsal (mobilization), the sympathetic (mobilization) nervous system, and the ventral vagus (social interaction and connection).

Perception: This is the term coined by Dr. Porges to explain how our autonomous nervous system reacts to the health, danger, and hazard of life from within our bodies, the world around us, and our relationships with the others. Like awareness, this is "detection without consciousness," a subcortical phenomenon that takes place far below the domain of conscious thinking.

Co-regulation: Polyvagal theory describes co-regulation as an evolutionary imperative: a need for survival. By controlling each other's autonomy, we feel safe to communicate and to develop trustworthy relationships. We should consider the autonomous nervous system as the basis for our living experience. This biological capital is the neuronal foundation below all knowledge. The way we travel across the globe— turning back, linking, and isolating at times — is directed by the autonomous nervous system. We are robust, enabled by co-regulatory partnerships. We are masters of living in the relationships of misadjustment. The autonomous nervous system "learns" the environment in each of our interactions and is toned for communicating or defending activities.

Hopefulness lies in understanding that the nervous system can be re-formed as early experiences mold it. Just as the brain continues to change in relation to events and the environment, our autonomous nervous system is also involved and can be affected intentionally. Because individual nervous systems reach out for touch and coregulation, resonance experiences and pain are perceived as bonding moments or defensive moments. The signs, health, or threat signals sent from one autonomous nervous system to another call for control or reactivity. In partnership with partners, one can quickly note the increasing reactivity when a conflict escalates rapidly, and the threats between the two nervous systems cause the need for safety for each partner. In comparison, the balancing of the therapeutic-customer relationship transmits safety signals and an independent demand for contact.

People are driven to want the "why" of behavior. We offer inspiration, motive, and responsibility. Society judges trauma patients in times of crisis through their acts. They also blame the victim too often because they didn't fight or try to escape but fell instead. We judge what somebody has done that leads to a trust in who they are. Trauma survivors frequently say, "It is my fault" and have a strong opponent inside who represents the reaction of society. We judge other people by how they communicate with us through our daily interactions with families, acquaintances, employers, and even informal conversations with strangers that define our day.

THE BASICS OF THE AUTONOMIC NERVOUS SYSTEM, INCLUDING ITS STRUCTURE, HOW IT WORKS, WHAT IT DOES, THE DIFFERENT NERVES

The autonomic nervous system is the portion of the peripheral nervous system that regulates the essential physiological mechanisms required to maintain normal physical functions. This functions without voluntary control, although certain events, such as pain, anxiety, sexual excitement, and changes in the sleep-wake cycle, alter the level of autonomy. The autonomous system is generally defined as the motor system, which is interconnected with three major tissue types: cardiac muscle, smooth muscle, and glands. Nevertheless, it also transmits visceral sensory information to and controls the central nervous system to modify the operation of specific autonomous motor outflows such as the ones which regulate the heart, blood vessels, and other visceral organs. It also induces the secretion of certain chemicals which are involved in the absorption of energy (e.g., leptin, glucagon, and epinephrine (a.k.a)) or cardiovascular functions (e.g., renin and vasopressin). Such synchronized responses sustain the body's normal inner atmosphere in a controlled state called homeostasis. There are two primary divisions: the sympathetic nervous system and the parasympathetic nervous system. Sometimes they work in antagonistic ways. Two serially associated neurons from the motor outflow of both systems. The second set, which is known as ganglion cells or post-ganglion neurons, is found outside the central nervous system in clusters of nerve cells known as autonomous ganglions. Parasympathetic ganglia tend to lie close to or in the tissues or organ that are intractable to your nerves, whereas sympathetic ganglia are found farther from your target organ. All mechanisms have related sensory fibers that give input on the functioning state of target tissues to the central nervous system.

Sympathetic nervous system

The sympathetic nervous system usually works to change the position of the cardiovascular system, such as sweat, as a response to a temperature increase. Nevertheless, the whole sympathetic nervous system is triggered under stress conditions and creates an urgent, widely-used reaction called the combat-or-flight response. This reaction is characterized by the release from the subcranial gland of large volumes of epinephrine, an increased cardiac production, skeletal muscle vasodilatation, cutaneous, and gastrointestinal vasoconstriction, pupil dilatation, and pilot dilation. The overall effect is to brace the individual for imminent risk. The dorsal horns of 12 thoraxes and the first 2 to 3 lumbar segments of the spinal cord have sympathetic preganglionic neurons. Axons of these neurons originate from the spinal cord in the ventral base, and synapse to either the sympathetic ganglion cells and advanced cells in the adrenal gland or chromaffin cells. The sympathetic network is sometimes referred to as the thoracolumbar outflow.

Sympathetic ganglia

Sympathetic ganglia can, depending on its location in the body, be classified into two main groups: paravertebral and prevertebral (or preaortic). Paravertebral ganglia are usually located on either side of the vertebrae and are connected to the sympathetic chain or trunk. Such ganglia typically have 21 to 22 pairs—3 in the cervix, 10 or 11 in the chest region, 4 in the lumbar region and 4 in the sacral area— and one unpaired ganglion, called the ganglion to impair, lies before the coccyx. The three cervical sympathetic groups (also known as the stellate ganglion) are the dominant cervical ganglion, the main cervical ganglion, and the cervical ganglion. The upper ganglion spans viscera of the head and innervates viscera of the back, thorax (i.e., bronchi and heart), and upper extremity of the central ganglia. The thoracic sympathetic ganglia enter the spinal area, and the pelvis and lower limbs are surrounded by the lumbar and the sacral sympathetic ganglia. The paravertebral ganglia all provide sympathetic innervation for muscle and skin blood vessels, hair-binding rectal pili, and sweat gland muscles.

The three ganglia preaortic are the celiac, upper mesenteric, and lower mesenteric. Located on the anterior surface of the aorta, preaortic ganglia develop axons that are spread by the aorta's three major gastrointestinal arteries. The celiac ganglion thus internalizes the heart, kidneys, pancreas and the duodenum, the first component of the small intestine; it internalizes the small intestine from the upper mesenteric ganglion; and it separates the bottom mesenteric ganglion from the lower colon, sigmoid colon, rectum, urinary bladder, and the sexual organs.

Neurotransmitters and receptors

When the blood vessels that feed them reach their target bodies, sympathetic fibers are done as a series of swellings near to the end bone. Thanks to this physiological structure, autonomous communication is carried out through a junction and not through a synapse. "Presynaptic" locations may be known because they include synaptic vesicles and membrane thickenings; on the other hand, post junction membranes seldom have morphological specializations but include certain receptors for a number of neurotransmitters. In contrast with standard synapses, the difference between pre- and post-synaptic components can be quite high. The difference between cell membranes, for example, of a standard chemical synapse, is 30–50 nanometers, whereas the gaps in blood vessels often go beyond 100 nanometers or, in some cases, 1–2 micrometers (1,000–2,000 nanometers). Due to these relatively large distances between the autonomous nerve terminals and their effector cells, neurotransmitters tend to act gradually. Some effector cells, such as those in smooth and cardiac muscles, are bound by resistance pathways that enable electrotonic cell coupling to compensate for this inefficiency. It allows many cells to respond and operate as a group if only one cell is activated. In the first example, the chemistry transmit the sympathetic mechanism appears simple: the pre-ganglionic neurons use acetylcholine as a neurotransmitter, whereas the rest of the post-ganglionic neurons use norepinephrine (noradrenaline). Nonetheless, neurotransmission is shown to be more complex on closer inspection as several chemicals are released, and each acts as a specific chemical code impacting various receptors on the target cell. Furthermore, these chemical codes act on presynaptic receptors found on their own axon terminals. they are self-regulatory.

The chemical codes are tissue-specific. Of starters, most protective blood vessel neurons secrete both NRA and neuropeptide Y. Major neurons that innervate the neural submucosal plexus of the gut possess norepinephrine and somatostatin. Sympathetic neurons that include the internal sweat glands have gene-related peptides, vasoactive intestinal polypeptides, and acetylcholine. However, besides the above-mentioned neuropeptides, certain molecules, such as classical neurotransmitters, norepinephrine, and acetylcholine, are produced from autonomous neurons. For example, certain neurons synthesize a gas nitric oxide that acts as a molecule of neuronal messengers. Therefore, sensory communication in the autonomic nervous system requires the activation of combinations of multiple neurosynaptic receptor agents.

Specialized macromolecules found in the cell membrane, neurotransmitters liberated from nerve terminals are linked to specific receptors. The binding activity initiates a sequence of complex biochemical reactions that produce a physiological response in the target cell. There are five groups of adrenergic receptors (receptors that bind epinephrine) in the sympathetic nervous system, for example, α1, α2, β1, β2, and β3. Such adrenoceptors are found in various concentrations throughout the body in different cells. Activation of α1-adrenoceptors in blood vessels causes blood-vessel constriction, thus activating the norepinephrine releases by α2 autoreceptors (receptors in a sympathetic pre-synaptic nerve ending). Different tissue forms have distinct adrenoceptors. Hearts and contractility are regulated by β1-adrenoceptors; β2-adrenoceptors are activated by bronchial smooth muscle relaxation; fat breakdown (lipolysis) is controlled by β3-adrenoceptors.

In the sympathetic system (as well as in the parasympathetic system), the cholinergic receptors (receptors that bind acetylcholine) are also located. Nicotinic cholinergic receptors induce the release of chemicals in the sympathetic postganglionic, adrenal chromaffin, and postganglionic parasympathetic neurons. Muscarinic receptors are primarily associated with parasympathetic processes of peripheral tissues (e.g., glands and smooth muscle). There are also peptidergic receptors in target cells.

The length of time each chemical operates on its target cell varies. Peptides usually cause long-lasting, slowly developing effects (one minute or more), whereas conventional transmitters create short-lasting effects (approximately 25 milliseconds).

Parasympathetic nervous system

The parasympathetic nervous system modulates visceral organs and glands mainly. Responses are never triggered en masse, as in a sympathetic response from battle to flight. While the parasympathetic system, unlike the sympathetic system, provides important regulation of many tissues, it is not necessary to maintain life.

The parasympathetic nervous system is similarly structured to the sympathetic nervous system. The engine portion consists of neurons of Preganglion and PG. The Preganglionic neurons are found within the brain stem and the lateral horns of the spinal cord at sacral levels in specific cell groups (also known as nuclei) (segmentsS2-S4). Preganglionic axons that arise from the brainstem to the parasympathetic ganglia found in the head (ciliary, pterygopalatine, and otic ganglia) or in the heart (cardiac ganglia), implanted in the end organ itself (e.g., trachea, otic ganglia) the body often names them craniosacral outflow). All pre-and post-ganglionic neurons are elusive for acetylcholine as a neurotransmitter but also include other neuroactive chemicals, which function as transmitters, including sympathetic ganglion cells.

The third cranial neuron (oculomotor nerve) includes nerve fibers that control the eye's iris and lens. From its origins in the Edinger-Westphal midbrain nucleus, Preganglionic axons fly into the orbit and synapse into the ciliary ganglion. The ciliary ganglion comprises two postganglionic neurons: one internal vates the iris smooth and causes pupil constriction, the other innervate the ciliary muscle, and regulates lens curvature.

Similar secretory glands are under parasympathetic regulation in the brain. These include the cranial gland that provides tears for the cornea of the eye; salvaguard glands that manufacture saliva and nasal mucous glands that secrete mucus in the nasal airways (sublingual, submandibular, and parotid glands). The parasympathetic preganglionic nerves controlling these functions are triggered by the reticular formation of the oblongata medulla. The superior salivary nucleus contains one community of parasympathetic Preganglionic neurons that lie in the rostral portion of the reticular formation of the médulla. Such nerves send out axons in the seventh cranial nerve (facial nerve) in the medulla in a separate branch of the intermediate nerve. Some of the axons are within the pterygopalatine ganglion, while others move towards the ganglion submandibular. The vasculature of the brain and head, lacrimal gland, nasal glands, and palate glands intervalley pterygopalatine ganglion cells, while the submandibular ganglion neurons interval submandibular and sublingual salivary glands. The second group of pre-ganglionic parasympathetic neurons belongs to the lower salivatory center in the caudal portion of the medullary reticular formation. Neurons of this category bring the ninth cranial (glossopharyngeal) nervous axons out of the medulla and into the otic ganglion. Post-ganglionic fibers migrate from this location to and inside the parotid salivary gland. The 10th cranial (vagus) nerve preganglionic parasympathetic fibers originate from two different sites on the medulla oblongata. Neurons that have a high heart rate originate from a part called the nucleus ambiguous in the ventral medulla, while the dorsal vagal nucleus is responsible for the functions of the gastrointestinal tract. Upon entering the vagus nerve medulla and heading to their respective tissues, the fibers synapse in the body's cells. Vagus nerve also includes the visceral afferent fiber that holds sensory information on a sensory nucleus situated in the medulla called the solitary tract nucleus, from neck organs (lynx, pharynx, and trachea), chest (heart and lungs) and gastrointestinal tract.

Enteric nervous system

The enteric nervous system consists of two plexuses, or neural networks, located in the gastrointestinal tract wall. The outermost plexus, between the inner circular and outer longitudinal smooth-muscle intestinal membranes, is known as the Auerbach or myenteric plexus. Peristaltic waves are controlling neurons of this plexus transfer the digestive products from oral to the anal opening. Therefore, myenteric neurons regulate local muscular contractions that induce stationary mixing and churning. Meissner or submucosal plexus is the innermost community of nerves. This plexus governs the luminous surface structure, monitors the secretions of the prostate, changes electrolyte and water distribution, and regulates the local flow of blood. Three endogenous enteric neuron function types are recognized: sensory neurons, interneurons, and motor neurons. Sensory nerves, stimulated by mechanical or chemical stimuli on the innermost surface of the gut, transmit information to interneurons inside the Auerbach and the Meissner Plexi and transmit information to motor neurons via the interneurons. The function of a number of target cells, including mucous glands, smooth muscle cells, endocrine cells, epithelial cells, and the blood vessel, is modulated by motive neurons.

The regulation of gastrointestinal functions also includes extrinsic neural pathways. There are three types: bladder, tactile, and electrical. Intestinofugeal neurons lie on the gut wall; their axons migrate to the ganglia and control reflex arcs covering large parts of the gastrointestinal tract. Sensory nerves relay distention and acidity information to the central nervous system. There are two types of sensory neurons: sympathetic neurons from dorsal-root ganglia at the thoracic and lumbar levels; and parasympathetic neurons from the nodose ganglion of the vagus or from sacral-level dorsal-root ganglia S2–S4. The first integrates the gastrointestinal tract from the pharynx into the left colic twisting, while the latter incorporates the distal colon and the rectum. Each part of the gastrointestinal tract receives a double sensory innervation: the pain impulses are transmitted by sympathetic afferent fibers, and the stimuli are detected by parasympathetic fibers to convey the information on the chemical condition of the stomach. The third extractive phase, which exercises motor control over the intestines, consists of parasympathetic Preganglion neurons located in the dorsal vagal nucleus of the medulla oblongata and the sympathetic Preganglionic neurons of the side horns of the spinal cord. Such pathways provide modular commands for the intrinsic enteric motor system and are not necessary for the absence of basic functions. The parasympathetic mechanism stimulates digestive processes through the mechanisms mentioned above, while the sympathetic system blocks them. The sympathetic mechanism impedes digestive functions by two means: (1) contraction of circular and smooth muscle sphincters found in the distal part of the stomach (pyloric sphincter), of the small intestine and of the rectum (internal anal sphincters), which serve as seals to inhibit the oral and reverse passage of digestive items. The parasympathetic organ, on the other hand, only sends signals to myenteric neurons.

Functions Of The Human Nervous System

In the great expansion and development of the cerebral hemisphere, the human nervous system is different from that of other mammals. Much of the roles of the human brain are understood from studies of disease effects, experiment results on animals, particularly monkeys, and animal and human health research in neuro-imagery. Such research has helped to explain facets of the neural system that underlie many properties of the human brain, including vision, memory, expression, and emotional processes. While scientists have quickly developed their knowledge of the functions of this unique system, it is far from complete.

To explain the workings of the human nervous system, scientists must first define the binding components or pathways between their various sections. Our research has led them to uncover neural pathways and establish less well-defined connections between different parts of the brain and spinal cord. It wasn't a simple matter to define such roads, and in fact, a lot of people remain unknown or are merely conjectural. A lot of information was gained on the human nervous system by studying the results of axonal disruption. If a nerve fiber is seven, the length of the axon, which is most separated from the cell, or soma, is robbed of and starts to kill the axonal flow of metabolites. The myelin sheath will also degenerate so that the degradation components of myelin can be seen with different stains under the microscope for some months after the accident. This technique is clearly of limited use in humans because correct injuries and subsequent evaluation are needed before the myelin is completely removed.

Degenerated axons and terminals, which form synapses with other neurons, may also be stained by using silver impregnation, but methods are laborious and sometimes difficult to interpret. It is not uncommon that a weakened neuron should have degenerative effects, albeit difficult to diagnose. However, neuronal degeneration is also shown to be interdependent. Neurons without sufficient input from damaged axons can atrophy themselves. This phenomenon is called the degeneration of anterograde. Similar changes may occur in neurons that have lost the main beneficiary of their outflow in retrograde degeneration. Such surgical approaches also refer to human disease. These can also be used post mortem if the central nervous system has been intentionally lesioned, for example, in treating intractable pain. Other methods can only be used in animal experiments, but these are not necessarily human-related. For example, natural biochemical components with a radioactive isotope may be inserted into neurons and carried along the length of the axon, where the radioactivity can be observed on an X-ray plate. The retrograde axonal flow measurement technique was used widely to show the sources of fiber tracts. In this process, the enzyme peroxidase is obtained by axonic terminals and brought to the soma by the axon, where the correct staining can be shown.

Neurotransmitter compounds can be spotted on both human and animal postmortem products. Nevertheless, success depends on the testing of fresh or frozen tissue, and the outcomes of the prior diagnosis with neurologically active medicines can be greatly affected. In an area of the nervous system, electrical stimulation induces nerve impulses in centers that transmit feedback from the stimulus source. This procedure, using microelectrodes, has been widely used in animal studies; however, it is difficult to identify the exact direction that the artificially generated impulses take. Many highly specialized imaging techniques, such as CT, MRI, and positron emission tomography, have given scientists the ability to image and study, in living healthy individuals, the anatomy and function of the nervous system. A technique known as functional MRI enables the identification of blood flow increases in tandem with brain activity changes. Functional MRI helps researchers to construct detailed maps of brain areas affecting human mental health and disease development. This approach was used in the study of different brain functions, from key visual responses to executive tasks.

Receptors

Receptors are biological transductors that convert energy in electric impulses, from both external and internal conditions. These may be massaged to form a sensory organ, such as the eye or ear, or scattered, like the skin and viscera. Receptors are linked by afferent nerve fibers to the central nervous system. It is called the receptive field, the region or location in the periphery from which a neuron receives feedback in the central nervous system. Receptive fields and not set agents are evolving. Receptors are of various varieties and are, in many respects, graded. For example, stationary state receptors produce impulses as long as a certain condition such as temperature remains constant. Changing state receptors, on the other hand, respond to changes in stimulus intensity or location. Receptors are also categorized as visually sensitive (reporting of the external environment), interceptive (sampling the body's environment), and proprioceptive (sensing of the body's position and motions). Exteroceptors record sight, vision, scent, taste, and touch senses. Interoceptors monitor the status of the liver, the food canal, blood pressure, and blood plasma osmotic pressure. Proprioceptors monitor the location and motions of the body parts and the body's place in space. Receptors are also categorized by the type of stimulus they are responsive to. Chemical receptors or chemoreceptors are sensitive to chemicals that are carried into the mouth or inhaled by the nose (smell or olfactory receptors) or which are present within the body (glucose or blood-based detectors). Surface receptors are graded as thermoreceptors, mechanoreceptors, and nociceptors, last sensitive to noxious stimuli or disruption to the tissues of the body.

Movement

The harmonious contraction and relaxing of chosen muscles contribute to body movements.

Contraction happens as nerve impulses are sent to the membrane protected by every muscle fiber via neuromuscular junctions. Most muscles do not contract constantly but are kept ready to contract in a state. The slightest gesture or even desire to move contributes to the extensive movement of the trunk and limbs ' muscles. Movement can be integral to the body itself and done by trunk and cavity muscles. Types include coughing, chewing, crying, sneezing, urination, and defecation. These movements are carried out primarily by the smooth muscles of the viscera (for example, the feeding channel and bladder), interspersed with ephemeral sympathetic and parasympathetic nerves. Many gestures connect the body to the world, whether to shift or signs to other individuals. These are done by the spine and limbs ' skeletal muscles. Skeletal muscles are connected to the bones and trigger joint movement. Efferent motor nerves and sometimes powerful sympathetic and parasympathetic nerves instill them in them. Each body movement must be right for strength, pace, and location. Such dimensions of movement are continuously identified by the body's location, stance, coordination, and internal factors sensors to the central nervous system. Such receptors are called proprioceptors, and those that comment on the location of the limbs constantly are the muscle spindles and tendon organs.

Sensory receptors

Only a subset of the nerve fibers that feed a muscle is the normal motor fibers. The others are afferent sensory fibers that inform the central nervous system of how the muscles are doing, or motor fibers that control the sensory nervous finishing behavior. If there is a continuous input of proprioceptive knowledge from muscles, tendons, and joints, there can still be an activity, but it is not possible to adjust it according to changing conditions or develop new motor skills. As mentioned above, muscle spindles and tendon organs are the major sensory receptors associated with body motion. The muscle spindle is much more complex than the tendon organ and is, therefore, less well known, although it was researched much more intensively.

Tendon organs

The tendon organ is simply a fibrous nerve afferent that ends with tendon slips in a number of branches where the tendons are attached to muscle fibers. The tendon organ is well-positioned to signal muscle tension by lying in series with skin. The afferent fiber of the tendon system is also sufficiently sensitive to provide a valuable signal that a single muscle fiber contracts. The tendon organs thus provide a continuous flow of muscle contraction information.

Muscle spindles

The common knee shock, regularly studied by physicians, is a spinal reflex in which a brief, rapid touch on the knee activates the muscle shock of an afferent neuron, which then triggers the motor neurons of the stretched muscle via a single synapse of the spinal cord. In this basic reflection, which is not conveyed by spinal cord interneurons, delays (about 0.02 seconds) occur mostly when the pulse is transported to and from the spinal cord.

HOW THE STATE OF THE NERVOUS SYSTEM AFFECTS DIFFERENT ORGANS AND ORGAN SYSTEMS

All processes in the bodywork to maintain a healthy environment. Every device has specific functions, but all are interconnected and dependent. The nervous system directly controls the different organs of the body. The brain frequently receives information from many bodies and changes feedback to ensure that they function properly. -structure and function of the human body are not regulated by nerves, although they play a major role.

There are three primary ways of controlling body organs and functions:

- ✓ Through the central nervous system
- ✓ Through the endocrine system
- ✓ Through local self-regulation (which includes intracrine, autocrine, paracrine, and immune regulation)

Here are some body organs and how they work with the nervous system

✓ **Skeletal System**

The skeletal structure shapes the body's foundation, which helps us to move as our muscles contract. It collects and releases minerals (for example, calcium, phosphorus) into the body as necessary. The nervous structure also protects internal organs and produces blood cells. Bones contain calcium that is essential to the nervous system's proper functioning. The cranium defends the brain from damage. The vertebras defend the backbone from damage. Sensory receptors in the joints of bones relay body orientation signals to the brain. Through regulating muscles, the brain controls the location of the bones.

✓ **Cardiovascular System**

The cardiovascular system pumps blood into the body to provide oxygen, hormones, nutrients, and white blood cells and eliminates waste products. The blood-brain barrier is formed by endothelial cells. Baroreceptors relay blood pressure information to the brain. Cerebrospinal fluid flows into the blood supply of the venous system. The brain controls the heart rate and blood pressure.

✓ Muscular System

Different muscle types require mobility, generate heat in order to maintain body temperature, transfer food through the digestive tract, and contract the heart. Muscle receptors provide the brain with body position and movement information. The brain controls skeletal muscle movement. The nervous system regulates the amount of food flowing through the digestive tract.

✓ Lymphatic System

The lymph system prevents inflammation in the body. The brain can trigger infection defense mechanisms.

✓ Respiratory System

The respiratory system provides blood oxygen and extracts carbon dioxide. The brain controls the respiratory rate and level of blood oxygen. The brain controls the breathing rate.

✓ **Digestive System**

The digestive system collects and digests fat, transfer nutrients to the bloodstream, waste, and water absorption. Digestive mechanisms provide some neurotransmitters with building blocks. The autonomous nervous system regulates the digestive tract sound. The brain controls the actions of drinking and eating. The brain controls the feeding and processing of tissues. The digestive system transmits sensory data to the cortex.

✓ **Reproductive System**

The reproductive system progresses to a new life. Hormones that replicate affect brain growth and sexual behavior. The brain controls the actions of the couple.

✓ **Urinary System**

The urinary system extracts waste products and ensures the equilibrium of water and chemical content. The bladder transfers sensory data to the brain .The brain regulates urination.

✓ Integumentary System

The integumentary system reduces water loss, includes receptors that respond to touch, control the temperature of the body, and prevent damage to the interior of the body. Body receptors transmit sensory data to the cortex. The integrated nervous system controls the flow of peripheral blood and sweat glands. Muscles associated with hair follicles are regulated by nerves.

Nerves bear instructions in the form of electrical signals from the brain and spinal cord. Nerves also allow us to perceive the tissue condition and transfer this information back to the brain and spinal cord so that we feel pain, joy, temperature, vision, hearing, and other senses. The body uses electrical nerve signals to regulate certain processes, and electrical signals can pass very rapidly. At the end of the axon terminal of each nerve, the electrical signals are translated into chemical signals that then activate the proper tissue response. The nervous system regulates, however, eventually the brain and the spinal cord, not the nerves, that only move through the signals. Some impulses are absorbed in the hippocampus, but dangerous signals are transmitted and responded through the spinal cord before the brain achieves the result we call "reflexes." The central nervous system plays an important role in regulating the body, but it is not the only organ that exercises control.

The endocrine system is a collection of endocrine glands that excrete various chemical signals, such as hormones, into the bloodstream throughout the body. The blood then circulates throughout the whole body, where various tissues usually respond to the hormones. An organ or system's response to a hormone depends on the amount of the hormone in the blood. It helps endocrine glands to regulate various organs and functions of the body by changing the amount of hormone they release. By contrast to the central nervous system, the control pathway is primarily molecular and not electrical for the endocrine system. The thyroid gland in the abdomen, for example, regulates the accelerated use of energy by secreting different levels of thyroid hormone in the body. Too much thyroid hormone, which leaves you nervous, jittery, and sleepless. Too little thyroid hormone and resting, lethargic and straight-thinking. A healthy body continually tracks the level of activity and regulates thyroid hormone levels when appropriate.

The adrenal glands which prepare the body to face an emergency and reproductive glands, which regulate body weight and reproduction, are also examples of endocrine glands. Hormones in body management work as diverse as libido, reproduction, menstruation, ovulation, abortion, breastfeeding, lactation, sleep, blood pressure, immune system blood sugar, vertical children's development, muscle mass, wound healing, mineral levels, appetite, and digestion. Essentially, much of the endocrine system is hypothalamic for the brain, but the endocrine system operates independently through feedback loops.

Eventually, the structures and activities of the body are regulated by local self-regulation. Instead of relying on the brain, organs and cells may do much on their own so that the brain is freed up for more important tasks. The organ can use distributed chemical signals, including paracrine hormone signaling, to transmit regulatory signals within the cell. Usually, these hormones are not transferred to the bloodstream but actually pass in between the cells. This approach works as paracrine hormones only work in adjacent cells. For example, blood clotting and wound healing are locally regulated by an exchange of paracrine hormones. Perhaps the liver is the organ with the greatest level of self-regulation. The liver moves well and performs hundreds of tasks from the rest of the body in one go. An organ can also interact electrochemically through its body. For, e.g., the heart does not pound, because it is informed by a nerve. The heart beats itself by a cyclical electric pulse stream. Although it is possible that the brain will order the heart to accelerate or slow down, the real heart rhythm is locally regulated. The growing cell of the body also has a certain degree of internal self-regulation within the cell itself. Many cells regulate themselves more consciously than others. For example, white blood cells are very active in hunting down and killing germs, as if they were individual entities. Active white blood cells do not wait until a brain or hormone orders them to function. Sperm cells are so independent that after they have left the male body, they can live and function properly. In addition, the central nervous system, endocrine system, and local regulatory systems are not autonomous but have complex influences over each other.

THE EXACT REASON WHY THE VAGUS NERVE IS SO IMPORTANT

Unlike the other Casino, it doesn't linger there in this vagus. The vagus nerve is a long meandering bundle of motor and sensory fibers, linking the brain to the heart, pulmonary, and intestines. It also branches into the heart, spleen, lung, ureter, female fertility glands, arms, mouth, tongue, and kidneys to touch and to communicate with it. It enhances our unwitting nerve core— the parasympathetic nervous system— which regulates the involuntary processes of the body, from steady heart rate and food intake and respiration and swelling. It also helps to regulate blood pressure and blood glucose level, supports the overall function of the kidneys, helps release bile and testosterone, enhances saliva production, helps control taste and tears development, and plays a major role in female fertility and orgasms. The vagus nerve has fibers that internalize nearly all our internal organs. Emotion control and diagnosis take place through the vagal nerve between the head, brain, and stomach, which is why we react strongly to extreme mental and emotional symptoms in our bodies. Vagus nerve dysfunction may result in a whole series of problems including hypertension, bradycardia (abnormally slow heartbeat), trouble in swallowing, gastrointestinal disorders, fainting, mood disorders, B12 deficiency,

In the meantime, vagus nerve stimulation has shown improvement in cases like:

- ✓ Anxiety disorder
- ✓ Heart disease
- ✓ Tinnitus
- ✓ Obesity
- ✓ Alcohol addiction
- ✓ Migraines
- ✓ Alzheimer's
- ✓ Leaky gut
- ✓ Bad blood circulation
- ✓ Mood disorder
- ✓ Cancer

A Closer Look At This Super Nerve

The largest of our 12 cranial nerves is the vagus nerve. Only the heart is a larger nervous system. Approximately 80% of its nervous fibers, or four of its five' lanes,' drive input from the body to the brain. The fifth lane travels in the opposite direction and passes impulses across the body from the cortex. Anchored in the brain stem, the vagus is dividing into the left vagus and the right vagus through the neck and arms. Through path consists of tens of thousands of nerve fibers that branch into your heart, lungs, stomachs, pancreas, and almost every other organ in your abdomen. The vagus nerve contains acetylcholine, a neurotransmitter that induces muscle contraction in the parasympathetic nervous system. A neurotransmitter is a type of chemical messenger released at the end of a nerve fiber that can transfer signals from one point to another that activates specific organ systems. Of starters, we would stop breathing if our brain could not reach our diaphragm through the release of acetylcholine from the vagus nerve. Many substances, like botox and heavy metal mercury, can interfere with the development of acetylcholine. The vagus nerve, which causes death, is believed to shut down Botox. Mercury prevents acetylcholine activity. When mercury is bound to the thiol protein in the cardiac muscle receivers, the cardiac muscle can not receive electrical impulses for contraction from the vagus nerve. Usually, cardiovascular complications priced. Mercury used in brain fillings with only 3,000 tons of mercury thrown into the atmosphere will interfere with the development of acetylcholine. Mercury-charged vaccines can also play a role in children's nerve autism. Diabetes, obesity, upper respiratory viral infections, or having part of the nerve sewn inadvertently during surgery can also cause vagus nerve injury. Stress, exhaustion, and fear will inflame the nerve. Even such an easy and weak pose can have a negative effect on the vagus nerve. Diet has also been claimed to play a role in the health of the vagus nerves. A high fat, high-carbon' cafeteria diet ' decreases the vulnerability of the vagus nerve. Spicy foods can also confuse them. This is because the enteric nervous system (ENS), which controls the operation of the gastrointestinal tract, interacts via the vagus nerve with the central nervous system. This is considered the digestive brain line. The ENS is sometimes referred to in our solar plexus as the second brain or replacement brain. The good maintenance of the stomach and vagus nerve portal affects our mental health. A recent study shows how antibiotics can make us aggressive if they disrupt our gut's microbiome equilibrium. A major study carried out in Hamilton, Ontario, Canada by McMaster University last year showed that certain helpful positive microbes could indeed reduce PTSD. According to a study by the National Center for Biotechnology Information (NCBI), probiotics may help maintain gut and vagal nerve signals in a healthy state.

Boosting With Electricity

Physicians used the power of the nerve on the brain for a long time. Vagus nerve stimulation, referred to as VAS, is sometimes used to treat people with epilepsy or depression. VNS is intended to prevent epilepsy by periodically transmitting electrical energy signals through the vagus nerve to the brain. Such bursts are given by a system like a pacemaker. It is situated below the skin on the chest wall, and a wire travels from it to the vagus nerve in the heart. Studies who studied the effects of vagus stimulation on epilepsy found that patients had a second advantage due to reducing seizures: moods have changed.

THE SURPRISING TRUTH ABOUT HOW YOUR LIFESTYLE CAN HARM THE VAGUS NERVE

Scientists have long recognized that chronic conditions, such as alcoholism and diabetes, can harm nerves, including the vagus nerve, but it is not well understood why that damage occurs. Neuropathy in many nerves can occur in people with insulin-dependent diabetes. If the nerve of the vagus is impaired, it may result in nausea, bloating, vomiting, and gastroparesis, which is too late to drain the stomach. Unfortunately, according to some experts, diabetic neuropathy can not be reversed. If the vaginal nerve is weakened by physical trauma or tumor growth, it can induce intestinal problems or hoarseness, vocal cord paralyzes, and cardiac delay. Many cases have been reported in patients whose vagus nerve damage has been sufficiently small that the nerve can regenerate once a tumor is removed, including a case published in Neurology in 2011.

The nervous system is a complex network that is highly specialized. This arranges, describes, and guides relationships between you and the world. Command of the nervous system:

- ✓ sight, sound, taste, scent, and feeling (sensation).
- ✓ Works voluntarily and accidental, including breathing, balance, and coordination. The nervous system also controls the functions of most other processes, including blood circulation and blood pressure.
- ✓ Ability to think and understand. Ability to think and justify. The nervous system allows you to be conscious and to have feelings, memories, and words.

The nervous system is broken down into both the cortex and the spinal cord (CNS) and the nerve cells (PNS) that regulate the voluntary and involuntary activity.

The signs of a concern with the nervous system depend on the location of the nervous system and the cause of the problem. Nervous system disorders will gradually arise and induce a progressive (degenerative) loss of function. Or it can happen suddenly and cause life-threatening (acute) complications. Mild to serious signs may be present. Some serious conditions, illnesses, and accidents that can cause problems in the nervous system include:

- ✓ Blood supply problems (vascular disorders).
- ✓ Injuries (trauma), especially injuries to the head and spinal cord.
- ✓ Problems that are present at birth (congenital).
- ✓ Mental health problems, such as anxiety disorders, depression, or psychosis.
- ✓ Exposure to toxins, such as carbon monoxide, arsenic, or lead.
- ✓ Problems that cause a gradual loss of function (degenerative). Examples include:
- ✓ Parkinson's disease.
- ✓ Multiple sclerosis (MS).
- ✓ Amyotrophic lateral sclerosis (ALS).
- ✓ Alzheimer's disease.
- ✓ Huntington's disease.
- ✓ Peripheral neuropathies.
- ✓ Infections. These may occur in the:
- ✓ Brain (encephalitis or abscesses).

- ✓ The membrane surrounding the brain and spinal cord (meningitis).
- ✓ Overuse of or withdrawal from prescription and nonprescription medicines, illegal drugs, or alcohol.
- ✓ A brain tumor.
- ✓ Organ system failure. Examples include:
- ✓ Respiratory failure.
- ✓ Heart failure.
- ✓ Liver failure (hepatic encephalopathy).
- ✓ Kidney failure (uremia).

Other conditions. Some examples include:

- ✓ Thyroid dysfunction (overactive or underactive thyroid).
- ✓ High blood sugar (diabetes) or low blood sugar (hypoglycemia).
- ✓ Electrolyte problems.
- ✓ Nutritional deficiencies, such as vitamin B1 (thiamine) or vitamin B12 deficiency.
- ✓ Guillain-Barré syndrome.

A sudden (acute) system problem, depending on the area of the nervous system affected, can cause several different symptoms. A common examples of acute problems are stroke and transient ischemic attacks (TIA). You might encounter one or more symptoms suddenly, such as:

- ✓ Naivety, tingling, fatigue, or incapacity to move part or all of the body (paralysis).
- ✓ Dimness, dimness, blurred eye, or visual loss in one or both eyes.
- ✓ Loss of speech, discomfort, or difficulty understanding speech.
- ✓ Immediately, serious tumult.
- ✓ Dizziness, exhaustion, or failure to stand or walk, particularly if there are other signs.
- ✓ Confusion or a shift in perception or actions.
- ✓ High nausea or vomiting.

Conflicts can also lead to sudden changes in mood, thinking, attitude, or perception. Abnormal body movements may or may not be present, such as twitching of the muscle. The frequency and severity of the seizures depend on the origin of the seizures and the area of the brain affected. Because of peripheral neuropathy or stroke, diabetes may cause balance problems.

Vertigo and dizziness are equilibrium (equilibrium) problems. Vertigo is often attributed to a medication or an inner ear or brain infection. Emotional distress, fatigue, issues with blood pressure, and other conditions can all lead to feelings of dizziness. Most headaches are not caused by serious problems with the central nervous system. The discomfort caused by a headache will range from throbbing or stabbing pain like migraine to severe pain, such as with cluster headaches for several days. Headaches are usually caused by sinus problems, scalp or head muscles.

Lifestyle Habits That Can Harm Your Vagus Nerve

➤ Poor Sleep

Sleep is vital to our well-being and particularly important for our brain's wellbeing. The brain replaces itself while we sleep. Lack of sleep quality inhibits this process, which, over time, makes the brain susceptible to damage. Americans are now less than ever awake. According to a CDC survey, more than a third of people report sleeping under 7 hours a night. Many individuals require 7 to 8.5 hours of sleep a day to ensure their overall health is optimum. Many of us have a poorer cognitive function due to poor sleep at night. The symptoms of poor sleep, sadly, go to reduced concentration and a weak personality. Recent studies have shown that poor sleep also causes brain cell degeneration and loss. That said, it is far too easy to sleep away in the hopes of increased productivity. Interestingly, increased sleep can improve productivity and reduce the time taken for completing tasks.

➤ Lack Of Personal Interaction

Speaking is incredibly helpful to your brain. The method of organizing and translating thoughts and feelings through vocabulary and making sense of words from the person or people you speak to is an excellent exercise for your brain. Research by the University of Michigan showed that memory and attention increased as little as 10 minutes a day. Higher levels of social interaction were also established as leading to higher cognitive function. Such findings were clear across all age groups. We now live in a time of more digital text or internet browsing replacing face-to-face experiences. The absence of real personal interaction not only limits your brain's' exercise' opportunities but also leads to higher rates of loneliness and depression, which contribute considerably to reduced brain health.

➢ Nicotine Consumption

As we mature, the brain cortex grows weak. This brain area is where essential aspects of learning, such as memory, vocabulary, and perception, take place. Nicotine increases the brain cortex thinning and may lead to Alzheimer's growth. While quitting smoking leads to cortical recovery, it is sluggish and may be incomplete. Basically, your brain is awful for smoking and vaping.

➢ Overindulging

Eating habits that are considered good for your body are good for your brain too. The abundance of salt, caffeine, alcohol, or food, in general, can affect the brain's health. Evidence from a 2012 study showed that overweight people had a decrease in cognitive function of 22 percent compared to slimmer people over 10 years. Over-indulgence sometimes does not cause lasting harm. However, over-indulgence lifestyle ends up jeopardizing all aspects of your health. Increased caloric intake and food choices are both linked to increased risk of cognitive impairment and Alzheimer's disease, according to Gad Marshall's behavioral neurologist.

- ✓ **Salt:** A contributing factor to high blood pressure is high salt intake. High blood pressure increases the risk of stroke, allowing the brain to undergo frequent mild damage, leading to slight cognitive impairment.
- ✓ **Sugar:** It's awful for your brain, as sweet as it can be. Studies reiterate that sugar has a negative influence on the brain structure and function, jeopardizing both long-term memory development and learning. A 2011 research has shown that sugar consumption is strongly linked to reduced cognitive function.
- ✓ **Alcohol:** A study published by the British Medical Journal found that people who ingest small doses of alcohol regularly (15-20 drinks a week) are 3 times more likely to cause perception

and spatial navigation problems in the brain area. The good news, unlike nicotine, is that the brain will cure itself and return to normal for those who stop drinking alcohol.

> **Being Sedentary**

Regular workouts are beneficial for your overall health. Research shows the benefits of human brain exercise. Aerobic exercise, such as cycling, gardening or running and weight lifting, leads to the health of your brain as follows:

✓ Improving hippocampal disease, a part of the brain linked to memory and learning. Increasing neuroplastic which are the ability of your brain to change as you learn, and new experiences.
✓ Blood vessels are strengthening, leading to better blood flow, which can help to stop plaque build-up related to dementia.
✓ You will gain the best brain effects if you raise your exercise to 45–60 minutes on most days.

✓ Whatever the number of days or the amount of time you spend exercising, it will make your body and brain safer.

It is easy to forget how important it is to look after your head, with so much attention paid to physical health. The health of the brain depends on a number of factors, including biology. While you can't control other things, you have control over lifestyle choices that can boost your overall brain health. Your brain is the most valuable thing; handle it well!

Lifestyle Changes Can Help With Nerve Pain

Changes in lifestyle alone probably won't make the nerve pain go away. We can help, though. Some few tips:

Stay Active

What does it really mean to stay active? It can be as simple as walking to the store, taking the stairs at work, leaving the car just far enough to take your rest and see you on the way to the shop. Do what you can to keep moving anywhere and anywhere. Try to sit up and down or move the legs to the front and back. If you feel comfortable, take a swim or aerobics exercise, both of which are excellent for minimizing nerve pain. Studies show that long-term pain patients suffer less when walking and have more stamina and improved moods.

Eat Well

Stay careful of the food you eat. Certain food sensitivities, including gluten, can exacerbate nerve sensitivity. It is strongly encouraged to participate in a gluten-free diet for patients with nerve pain. To eat 5 to 10 portions of bright fruit and vegetables every day. Try to avoid gluten-filled foods such as pasta and bread. No special diet is required for nerve pain. But you will have the nutrients that you need through a well-balanced diet–with lots of fruits, vegetables, and whole grains. Take into account restricting or reducing alcohol.

Laugh Often

It's not any easier. Laughter is the best medication for pain repair, memory recovery, and comic relief. According to Dr. Lee Berk, an assistant teacher at the Allied Health Professions College in Loma Linda, the humor found that in their 60s and 70s, they enhanced the memory of adults. Love the little moments of life.

Talk It Out

What do you feel today? Do not be afraid to talk to colleagues, a priest, or even a psychologist about your day or your thoughts. Talking is a great way to communicate good or bad feelings. It may sound as though a heavyweight is taken off our shoulders as we speak to others.

Manage Stress

Learn how to manage your stress every day. Take a minute to relax when you are tired at work or at home. Feel free to walk out and have a short walk. Don't feel like you have everything to handle. It's all right to ask for help. Chronic stress can lead to a variety of ongoing health issues, including depression, anxiety.

Don't smoke.

Tobacco decreases brain blood supply and exacerbates nerve pain.

HOW THE BODY REGULATES STRESS AND DEPRESSION AND SUREFIRE WAYS TO EXPEDITE THIS PROCESS

For a long time, experts have proposed that hormones have receptors only in the peripheral tissues and have no access to the CNS. Nevertheless, the impact of anti-inflammatory medication (which is called synthetic hormones) on mental and cognitive problems and the phenomenon known as "steroid psychosis" is shown through studies. Neuropeptides first recognized in the early sixties as molecules without effects on the peripheral endocrine system. It was, however, found that hormones can have biological effects on various parts of the CNS and play a significant role in behavior and cognition. In 1968, McEwen suggested for the first time that the brain of rodents could respond to glucocorticoid (as a stress cascade operator). This theory was accepted that stress might induce functional changes in the CNS. Since then, two forms of corticotropic receptors were identified (glucocorticosteroids and mineralocorticoids). It was found that cortisol and corticosterone are roughly one-tenth of the affinity of the glucocorticosteroid receptors for mineral corticosteroids. Both types of receptors are in the hippocampus area, whereas other brain areas have only glucocorticosteroid receptors.

For 50 years, the effects of tension on the nervous system have been studied. Other experiments have shown that stress has many impacts on the human nervous system and can induce structural changes in various brain areas. Chronic stress can lead to brain mass atrophy and weight reduction. These structural changes involve variations in pain, attention, and memory responses. Clearly, the magnitude and severity of the changes depend on the stress level and stress length. Nevertheless, stress is now apparent that structural changes in the brain will induce long-term nervous system damage.

Stress and Memory

Remembrance is one of the main functional aspects of CNS and is graded as tactile, short-term, and long-term. Short-term memory relies on the front and parietal lobe, respectively, while long-term memory depends on the activity of large areas of the brain. Nevertheless, total memory capacity and short-term memory transfer to long-term memory rely on a hippocampus, which has the highest density of glucocorticoid receptors and the lowest reaction frequency. Therefore the relationship between hippocampus and stress has been heatedly discussed over the last several decades. In 1968, cortisol receptors were proven in the hippocampus of rats. In 1982, the existence of these two receptors in the brain and hippocampus area of rats was demonstrated by the use of specific agonists of glucocorticosteroid and mineralocorticoid receptors. It is also important to note that the amygdala is important for assessing emotional memory memories.

The findings of past studies have shown the impact of tension on the memory cycle. Various studies have shown that stress in the hippocampus brain can induce functional and structural changes. Such structural changes include neurogenesis and atrophy diseases. In turn, the number of dendritic branches and the number of neurons, as well as the structural changes in synaptic terminals and reducing neurogenesis in hippocampus tissue, was reduced by chronic stress and, therefore, the increase in plasma cortisol. Glucocorticosteroids can induce this by either influencing cellular metabolism or increasing hippocampal cell sensitivity to stimulating amino acids (Sapolsky and Pulsinelli, 1985[98]) and/or raising extracellular glutamate levels.

High-stress hormone concentrations can cause reported memory disorders. Animal studies have demonstrated that stress can result in a reversible reduction in spatial memory due to hippocampal atrophy. Indeed, elevated plasma concentrations of glucocorticosteroids over long periods of time can contribute to memory deficits atrophy of the hippocampus. In fact, people who receive either Cushing syndrome (with elevated glucocorticosteroid secretions), or who receive high doses of exogenous prescription anti-inflammatory medications, are reported to have hippocampus atrophy and related memory problems. MRI pictures taken from the brains of people with post-traumatic stress disorder (PTSD) show a decrease in hippocampus volume and neurophysiological symptoms as well as a poor verbal memory. Many human studies have shown that even the standard therapeutic doses of glucocorticosteroids and dexamethasone can cause specific memory problems. There is, therefore, an inverse relationship between cortisol level and memory that raises plasma cortisol levels after prolonged stress contributes to a memory loss that occurs when the plasma cortisol level declines.

Stress also impacts performance adversely. Data from loading hippocampus-dependent data show that participants have not been subjected to a new environment too common. In addition, adrenal steroids lead to potential long-term modification (LTP) as an essential memory formation process.

Under pain, two factors are involved in the memory cycle. The first is noradrenaline, which produces emotional aspects of memory in the amygdalous basolateral region. Furthermore, corticosteroids promote this process. Furthermore, if a few hours earlier, the activation of corticosteroids triggers amygdala inhibition and subsequent behavior. There is thus a mutual equilibrium between these two hormones to create an answer in the memory process.

Stress does not affect memory at all times. Often stress can actually improve memory under special conditions. Such symptoms entail unfamiliarity, non-predictability, and life risks to the stimulus applied. Under these particular conditions, stress will temporarily enhance brain function and memory. In fact, stress may sharpen memory in certain situations. For example, it was shown that a written test would improve memory in study subjects for a short period of time. This disease is curiously associated with a drop in the cortisol intake of the saliva. Many experiments have shown that imminent stress before learning can also result in an increase in brainpower or a decrease in memory capacity. This paradox is the product of the type of stress that is introduced and either the magnitude of the emotional connection with the stressful event or the length between the stressful and the learning process.

The memory restoration cycle is normally improved after pain. Various studies in animal and human models have shown that either glucocorticosteroid administration of discomfort shortly after learning promotes recall. Glucocorticosteroids (not mineralocorticoids) have also been shown to be necessary to improve learning and memory. The memory recovery following stress consumption will, however, be limited, which can result from the competitiveness of updated data for memory storage in a stressed environment. Such a study has shown that either stress treatment or glucocorticosteroid injection before a consolidation test lowers the brain capacity for humans and rodents.

In short, it was found that the effect of stress on memory depends very much on the duration of the response to traumatic stimuli and memory can be stronger or worse when it comes to timing the stress. Recent studies have further shown that using a precise scheduled schedule for traumatic treatment influences not only hippocampus-dependent memory but also striatum-dependent memory.

The role of stress in the cause and progression of depression can be clarified by several converging causes, including the systemic effects of environmental stressors and the long-term effects of childhood stress events, which can trigger sustained hyperactivity on the hypothalamic-pituitary-adrenal axis. The shifts, including higher corticotropin release and cortisol production, are also related to amygdala hyperactivity, hippocampus hypoactivity, and reduced serotonergic neurotransmission, contributing in addition to enhanced susceptibility to stress. When creating a detailed model for the relationships between multiple risk causes, the functions of other monoaminergic neurotransmitters, genetic polymorphisms, epigenetic pathways, inflammatory processes, and impaired cognitive processing were also addressed. Further knowledge of the processes driving these causes will make a significant contribution to creating further effective treatments and prevention approaches on the connection of stress and mood disorders.

The connection between stressful life events and the origin and development of depression has been studied extensively and has provided an increasing number of evidence supporting this association. Environmental factors are likely to affect individuals in different ways, which leads to an adaptive response to stress that depends on the interaction between stressors and individual resources both on the psychological and biological aspects. All executive analysis related to incoming information involves psychological aspects; the subjective assessment of different aspects related to stressors, including severity and chronicity, predictability and controllability, and potential tools for coping with them. Biological mediators involve activation of various neuronal systems, including sensory receptors that relay external feedback to the CNS, as well as activation of neural and neuroendocrine cascades of biochemical events, and as expressed through the resulting stimulation of sympathetic control and hypothalamic-pituitary differentiation of the autonomous nervous system If it carries on in a lengthy and repetitive fashion (e.g., in chronic stressful situations) it can lead to maladaptive changes that in turn will help develop behavioral disorders such as anxiety and mood disorder, like depression, particularly in people with elevated genetic susceptibility. Different polymorphisms as candidate genes have been investigated, known to participate in important molecular pathways of depression origin. The presence of these genetic variations appears to be involved in the development of depression when stressful events such as childhood adverse events and adult environment stressors are involved. In addition, different studies focused on gene-environmental interactions, including the search for the polymorphic variants and the role of epigenetic transcriptional regulation. Moreover, inflammatory processes, combined with adaptive responses to stressful situations, with the consequent abnormal synthesis and release of proinflammatory cytokines, can lead to further maladaptive changes in neural and neuroendocrine systems and therefore help to develop depressive symptoms especially in individuals with chronic stress.

This chapter explores evidence of the role of stress in various converging factors such as genetic diathesis, history of detrimental early life experiences, HPA-axis hyperactivity, decreased monoamine, enhanced pro-inflammatory cytokines, and epigenetic pathways such as those found in response to environmentally stressful conditions, and their future stressful conditions. Increased understanding of these factors and their potential interactions can lead to more effective treatment strategies.

Processing of Environmental Stressors in the Brain

Environmental stressors are perceived and conveyed via sensory channels to various structures of the CNS, for instance, thalamus that send projections to the amygdala, sensory and cortical tissue and in turn to separate prefrontal cortex (PFC) regions, including the orbitofrontal cortex, medial PFC and anterior cingulate cortex (ACC). The direct projection from the thalamus to the amygdala helps to trigger arousal and early alarm reaction, as the autonomous nervous system and the HPA axis subsequently become activated, while indirect projections of sensorial and associational cortices and transition cortices can reach the amygdala. The latter regions, including the entorhinal, perirhinal, and parahippocampal cortices, are also intended for the hippocampus in which sensory input is combined with qualitative signals to provide the amygdala with more detailed information.

Throughout emotional processing, the amygdala plays an important role, namely determining the emotional significance of environmental stimuli and inner stressors. It plays a key role in the control of self-replications and neuroendocrine reactions, by projecting the lateral hypothalamus which mediate activation of the sympathetic branch of the autonomic nervous system, by direct projections into the paraventricular nucleus of the hypothalamus or by indirectly activating the bed nucleus of the stria terminalis Furthermore, the amygdala shares important connections to the orbitofrontal cortex and medial PFC, which include the Brodmann and subgenital ACC areas. The orbitofrontal cortex has been associated with the integration of the multimodal sensory stimuli and the primary evaluation of their positive or negative value. The Medial PFC overlaps with the ACC, particularly in the subgenuine ACC, which controls the amygdala emotional responses. In addition, these systems are related to the dorsolateral PFC and ventrolateral PFC that are involved in cognitive control and voluntary emotional regulation. The dorsolateral PFC, related to executive aspects of cognitive processing (especially the conscious management and working memory), receives feedback from the amygdala via the oropharyngeal and ACC. The dorsolateral PFC contributes to limbic systems, primarily by means of indirect connections to the ventromedial PFC and to the subgenual ACC. Projections from the ventromedial PFC and sub-genre ACC were suggested to modulate the amygdala, which, in effect, sends a thrilling contribution to the hypothalamus, thereby controlling the HPA axis function.

The decreased volume and hyperactivity of the ACC in people with mood disorders, which have been related to the function of the ACC in the top-down regulatory pathways between dorsolateral PFC and amygdala, have been identified to intentionally downregulate negative emotions. For patients with depression, these corticolimbic channels may be dysfunctional, and display impairment of dorsal PFC, dorsomedial PFC, orbitofrontal cortex and ACC, especially during cognitive-emotional work, resulting in a disturbance of their top-down effect inhibitory in cognitive-emotional impaired control. Conscious improvement of negative emotional control was related to psychological rehabilitation. In combination with elevated amygdala activation and diminished dorsolateral PFC function, decreased hippocampal volume was also observed.

Relating The Polyvagal Theory To Add

The "strong alert, vagus signs" theory suggests that people with Autism Spectrum Disorder (ASD) face multiple problems. The theory says that spectrum people have reduced their ability to engage with the world because they have not learned to process complex social data. As a result, they will continue to read environmental risks and quickly jump into Fight / Flight / Freeze and shut down. The hypothesis also suggests that when people grew up, the vagal function was redirected to body protection, and this became the subject of the growing child. When the body system is immobilized, it becomes painful or agitated, digestion is difficult, and the focus is directed toward the outside world that compromises interaction. This suggests that the normal development of the social engagement mechanism does not take place as it should, and thus the infant may not learn to use its system and end up as an "autopilot." The Polyvagal Theory is based on early response to stress and trauma, and for any cause, the vagus is ready for realignment, and the nervous system of the infant is not fully developed. It may be emotional trauma, or the terror in utero or at conception, or a physical obstruction of the vagal nerve. The environment of the kid is in a state of distress. Instead of working in harmony with the old and the new system, they end up sitting separately. In other words, the kid lacks all that the social system provides, and the senses are more or less offline. Polyvagal Theory suggests that autism is a learned response to early stressors–the result of a child being in a protracted state of' fight or flight.' The book explains the idea in simple terms and provides recent developments in brain plasticity science (the brain's ability to change in life) to give parents and clinicians the resources to improve the brain-body bond with children and to decrease the psychological and emotional impact of autism.

USING YOGA POSES AND STRETCHES TO HELP YOU ACTIVATE THE VAGAL NERVES

Persons with maximum vagal tone are more stress-tolerant and can switch from an excited state to a relaxed state and vice versa without unduly disrupting them. These people tend to have good resistance and are safer. A low vagal tone, by contrast, is associated with inflammation, emotional stress, negative moods, and heart attacks. People with low vagal tone have slight digestion and often have physical, mental, and emotional dysfunction. The ideal vagal tone is therefore related to physical and psychological well-being, while a weak vagal tone is connected to fatigue, depressive moods, stress, and cardiac problems.

Since the vagal tone determines the functioning state of the vagus, it may also appear clear that people with a preferably effectively vagal tone turn on the parasympathetic nervous system to "feel good and cool" as opposed to people with a low vagal tone who experience the nervous system activity and who are generally in tension or difficulty and in flight. The strength of the vagal tone can predict a better prognosis of health status, especially in people recovering from chronic disease. The vagal tone, however, is not static; it fluctuates depending on our habits, lifestyle, and moods. We must all aim to increase the vagal nerve function to maximum levels and to preserve the ideal sound. Today's report is about yoga activities that keep the vagal tone functioning optimally. The vagus nerve stimulates essentially the parasympathetic nervous system–the "rest and digesting" mechanism of the body that functions when the body is not stressed. Physicians often add slightly immune nerve stimulators in patients, but yoga can do the same.

How does Yoga stimulate the vagus nerve?

Asanas and Vagal Tone

As you perform asanas, the parasympathetic nervous system can be stimulated by massaging the lungs, enhancing breathing, muscle relaxation, and regeneration of the mind. Sarvangasana, for example. Sarvangasana, or the Shoulder Hold, often specifically activates the vagus nerve as it travels through the back. To optimize activation of the vagus nerve and secure it simultaneously, use two or three folded yoga blankets to support the pose. Place the folded blankets equally on each other and sit on the bottom of blankets with your head. It allows the covers to support your body weight as you step into the stance. The neck and vagus nerve are not excessively squeezed, and you can feel the triggering of the PNS posture. Gravity works to drive most of the body's blood into the uterus, where it can support muscles. It increases the capacity of your body to relax and eat. Since the vagus nerve penetrates the throat, activities, or actions that affect the flow of energy through the vagus nerve will have a profound effect on breathing and circulation. In tandem with air, yoga asanas activate both electric and mechanical movements in the body. Electrical activity is expressed in the vagus nerve stimulation that initiates the brain and goes directly to the organs that activate mechanical actions. The mechanical action is as follows:

✓ Contraction and relaxing of most, if not all, skeletal muscles in the body

- ✓ quick blood and lymph flow through the arteries
- ✓ the beating of the heart
- ✓ smooth venous return respectively

Yoga exercise also preserves the vagal tone and encourages a move from combat or flying to calming mode.

Smooth or complete inverts, a flowing stream that involves moving from Sun Salutations into each asana and soft backbones, both optimize vagal sounds, slowly warm up the body and the cardiovascular system, relaxing and expanding the connective tissues and muscles that line the abdomen to allow more efficient blood flowing.

Inverted Poses. Such positions using gravity to control blood pressure (the heart is equal to the head). Besides complete inversions, such as Headstand and Shoulderstand (Sarvangasana), a number of "partial" and "low" inversions are as effective in reducing stress as complete inversions (see All about supported investments). The legs up the wall pose (Viparita Karani), and the supported bridge poses (Setu Bandha Sarvangasana) are classic examples. The relief with these partial / smooth inversions is close to that obtained with restaurant poses.

Flow Sequences. Such stages involve moving from Sun Greetings to an asana. This cycle encourages endurance, strengthens the back, extends muscles, and allows blood flow smoother (Cultivating Endurance in Yoga). Flow sequences are invigorating and can be very stimulating, especially if you are slow, depressed, or lethargic (a sign of optimum vagal tone).

Backbends. These positions also contribute to the relief of leniency, dullness, and lethargy (a sign of vagal tone). The standing positions could be adjusted with the arms raised upward and transformed into a gentle backbend as in the mountain stance (Tadasana). Exhale from here, drop your arms down to the floor and continue your Sun greetings. Even a calming posture like Savasana could be adjusted to stabilize the torso with a bolster or stack of blankets to raise the chest and let the back of the torso be in a relaxed backbend.

The above cycles have a quiet impact on the nervous system, which facilitates a change from battle or flight mode to rest and digest mode, a trait for a vagus nerve that functions optimally.

Pranayama and Vagal Tone

While we do not have power over the autonomic nervous system, we can move from fight or flight to relax and digest by pranayama (breathing practices). Note, you should alter your breathing freely, and this is the secret to altering the flow of energy from the distinct nerve and to changing the nervous system. This is a tool that can be used in moments of acute stress. For example, one-second inhalation and two-second expiration should include a longer exhalation than inhalations in pranayama practice to calm down this stress response. Furthermore, smoking on exhalation as in Bhramari Pranayama or gradual exhalation in Sitali breath slows the rhythm of breathing. This reduced heart rate sends the brain a message that circumstances are more peaceful and stimulates the autonomic nervous system to calm / relax, recover, and heal in the Rest and Digest mode. Pranayama methods, which broaden exhalation, are therefore excellent for enhancing the vagal tone. If your inhalation and exhalation are the same lengths, for example, by intentionally taking a three-second inhalation and three-second exhalation, or by practicing alternative nose-borne breathing, the vagal tone of the breath is subtly influenced according to your current condition. This method is, therefore, effective at preserving the ideal vagal tone, which will stimulate the subconscious at the present time without having a strong effect on the nervous system. As we respire intensely and gradually, we stimulate the vagus nerve and activate the parasympathetic nervous system. The PNS is intended to turn the brain and nervous system off and helps to calm the processes. It promotes relaxation and rest by slowing down our heart rate, slowing down our breathing, constricting our eye pupils, decreasing muscles and relaxing the tissue in the whole body.

Mantra

The vagus nerve controls physical functions in the throat, larynx, and ears— the area called the chakra of the throat. Singing or listening to singing stimulates nerves in the throat through vibration and releases blocked energy.

The following distinctly nerve yoga activities will help you develop a good vagus rhythm, stimulate, calm, and stabilize your life:

Conscious respiration: breathing is the most effective way to change the equilibrium between the supportive and caring behavior of the nervous system. Vagus nerve yoga emphasizes on diaphragmatic relaxation and increasing the duration of the exhalation to counteract any over-stimulation of the positive nervous system. Evidence has shown that steady, rhythmic diaphragmic breathing improves vagal wellbeing. A type of yogic breath is Ujjayi pranayama, which, by shifting the whisper muscles, causes slight constriction in the back of the throat. Exhale out of your mouth and know this wind, like you are fogging a mirror. Now, breathe in the same way, close your mouth and exhale your nose. The sound of your breath is louder and often sounds like ocean waves. Start with a count even for your inhalation and exhalation. For even more relief, increase the length of your exhalation slowly relative to the inhalation. For e.g., you can start with a 4-count inhalation and exhale the exhalation into a 6 or8-count exhalation. It calms the parasympathetic nervous system.

Half-Smile: Engaging in a "half-smile" is a helpful way of improving your mental state and at least maintaining a serene feeling. As the vagus nerve reaches into the muscles of the neck, you can raise your vagus tone by relaxing your facial muscles and turning your lips slightly. This practice helps to involve the most evolved branch of the vagus nerve called the "social nervous system" by Dr. Stephen Porges. Imagine your jaw smoothing and a calm feeling running through your face, head, and shoulders. Note the fragile shifts in your perception and emotional consistency.

Open your heart: With yoga postures that close your mouth and arms, you can softly activate the subtly nerve. Consider this soft opening exercise with your palms on your back. Inhale your elbows high as you stretch across the front of your arms and raise your chin. Exhale your knees above your eyes and tuck your chin. In this moving meditation, take some deep breaths. The emphasis on your inhalation can be calming and relaxing in this breath pattern. Allow yourself to extend to the open heart.

Wake up and stretch: If you have a hard time waking up in the morning, or feel tired and lenient at night, yoga will pick me up for your mind and body. Explore the standing positions of such a militant (virabhadrasana) to affirm your mind and wake you up. Note that your feet are connected to the Earth to remain grounded to energize you in a balanced way. Let your breath remain rhythmic so that you stay grounded and connected to your body's sensations.

Release the belly: you can work through the gut by connecting the vagus nerve. Find your way to a table with your palms under your arms and your feet behind your thighs. If your knees are sore, you should put a folded pillow beneath you. When you inhale, you start lifting your head and hands to the floor as you step into the Cow Pose. You lower your head and hips on your exhalation while you lift your spine up to Cat Pose. Find your own movement rhythm with your wind. Repeat as often as you would like to create a gentle massage for your belly and spine.

Self-compassion and "loving-kindness" meditation: Self-compassion and the practice of "loving-kindness" urge you to devote yourself and others to the act of friendship. Research on individuals who practice loving-kindness has revealed a greater vagueness, greater autonomy, a greater sense of social relationship, and more positive emotions. Take a moment to think about a challenge that you face in your life. Now, imagine another person facing a similar challenge. Would you invoke this other person a sense of compassion or kindness? Note how the body feels this sense of love. I wish them well. I wish them well. See if you can extend to yourself this same quality of loving-kindness? I wish you well. I wish you well.

Yoga Nidra: Restaurant yoga will help slow down the nervous system and soothe it. Yoga Nidra is a classic practice, often referred to as' yogic sleep' or relaxing meditation. Yoga Nidra is the cure for our exhausting, modern lifestyle, and offers the chance, through part sympathy, to regenerate body and mind. When you reach a comfortable spot lying down on the floor, pillow or yoga mat, lift your body, and breathe consciousness. Make room for everything you experience, including any stress, heaviness, or constraint zones. Give yourself a truly soothing and nourishing feeling for 30 minutes.

PROVEN MEDITATIVE TECHNIQUES TO HELP YOU STIMULATE THE VAGAL NERVES

A good function of the vagal nerve is critical for optimal health. Emerging research suggests that many chronic disease states may not perform well. If your goal is to stimulate your vagus nerve to boost your mood or stress symptoms such as panic disorders and anxiety, it is important that you speak with your psychiatrist, particularly stress if your daily life has a significant effect. Major changes of mind, such as excessively depressed, depression, constant low mood, euphoria, or fear, are all excuses to see a doctor. Remember that there is no evidence that poor vaginal tone triggers anxiety or mood disorders. Complex disorders such as anxiety include a wide variety of different variables, including brain chemistry, the environment, health status, and biology. In comparison, changes in nerve tone and brain chemistry can not be modified independently with the methods mentioned above. The reasons listed in this book are instead designed to reduce daily stress and encourage optimal mental health and wellbeing. Most are backed only by small observations of humans or animals. You can, therefore, try the following methods if you and your doctor agree that they are suitable. Learn the methods we take and talk to your doctor before checking them. This is especially important if you plan on taking some dietary supplements. The FDA has not approved medical supplements and typically lacks strong clinical research. Regulations lay down for them production standards but do not promise that they are safe or effective.

Finally, remember that none of these strategies should ever be carried out in place of the recommendations or directions made by your doctor.

Cold

For one test on 10 healthy people, the fight-or-flight function reduces, and the rest-and-digest (parasympathetic), which mediates through the vagus nerve, improves if the body adjusts to cold temperatures. Temperatures below 50 ° F (10 ° C) were deemed cool in this analysis.

Sudden introduction to cold (39 °F/4 ° C) also increases the activation of vagus nerves in rats. Even if the effects of cold showers are not documented in the vagus nerve register, many favors this traditional method of cooling. When we think about it, all showers were cold showers before water heating techniques came into being. Cold tubs are anecdotally popular in Japan, while many northern nations dip in the ocean during winter and early spring for special occasions. Nonetheless, it usually takes a while to get used to cold showers. Some people say it's good to start by dipping your face in cold water. Please remember to first contact the healthcare provider. For people with or at risk of heart disease, most doctors recommend cold showers. This is because sudden cold exposure will restrict blood vessels that can increase blood pressure and heart rate.

Singing or Chanting

According to an interesting 18-year-old study, singing increases the rate of the Heart Rate (HRV). Relaxation, greater stress tolerance, and adaptation and higher rest and feeding (parasympathetic) behavior correlated with heart rate variability. The writers of the above study found that humming, prayer chanting, hymn singing, and vigorous singing all improve HRV in some way. We believed that singing initiates a vagal pump, sending calming vibrations through the chorus. Singing at the top of your lungs could also manipulate the muscles behind the throat to stimulate the vagus. The authors of this study, on the other hand, think that enthusiastic singing stimulates the sympathetic nervous system and the vagus nerve, which can help people float. Singing in unison, often done in churches and synagogues, has also increased the function of HRV and vagus in this study.

No other similar studies were, however, undertaken. The study mentioned above included only 15 stable 18-year-olds. We don't know how the vagus nerve of people of different ages or suffering from mental health problems impact various types of music and performing. Larger studies are necessary. In the only other research concerned with this connection, singing in both professional and amateur singers was found to increase oxytocin. After a singing session, both groups felt energized, but amateur singers felt better and more excited than professionals. The authors noted that this could be because amateurs treated singing as a technique of stimulation and self-realization, while the professionals were successful. So, as you listen and perform, you may want to relax and show yourself as much as possible. Don't try to think about how you are progressing, and if you are going to achieve the goals you set for that session.

Yoga

Few studies suggest a link between yoga and decreased vagus nerve function in general and parasympathetic system activity. 12-week yoga therapy was better related to mood and anxiety changes than the control group walking workouts. The study found increased levels of thalamic GABA associated with improved mood and reduced anxiety. Yoga is considered good for both physical and mental health care. Further work on its effects on the vagus nerve tone is needed.

Meditation

Research shows that at least three forms of meditation will partially activate the vagus nerve. In small studies, love meditation, gratitude meditation, and Om singing have been associated with increased variation of the heart rate. Many scientists believe that this influence could be underpinned by intentional, deep breathing that follows meditation and other contemplative activity. Careful breathing is thought to relax the vagus nerve and rest and absorb nerve activity directly. Further human studies and experiments on different types of meditation are required.

Positive Thoughts and Social Connection

In a study conducted by 65 men, half of the participants were told to sit and think with humility on others by silently repeating phrases such as "May you feel safe, may you feel happy, may you feel healthy, may you live easily." The meditators showed an increase overall in positive emotions like happiness, interest, amusement, serenity, and hope after the class in comparison with the controls. These emotional and psychological improvements were linked to a greater sense of connection with other people and an improved vagal function, as measured by cardiac variation. Nonetheless, actually meditating did not always lead to a more toned vagus nerve. The change took place only in meditators who were happy and socially connected. Those who meditated so much but did not feel similar to others did not show any change in the sound of the vagus nerve. While further research is needed, these findings indicate how the vagus nerve will help people on a journey towards better health. Positive thoughts and social links may stimulate the vague nerve and foster joy, peace, and compassion.

Deep and Slow Breathing

Slow and steady breathing is also thought to activate the vagus nerve, and various types of meditation, yoga, and calming methods are probably common. The nerves of your heart and neck produce "baroreceptors." These sensitive neurons sense stimulation in your blood and carry on the synaptic signal to the brain (NTS). This signal stimulates a person's vagus nerve, which connects the heart with decreased blood pressure and heart rate if a person's blood pressure is high. The effect is less (sympathetic) combat-or-flight activation and more relaxing and digesting (parasympathetic) activity. Baroreceptors can be prone to vary degrees. Many researchers think the more alert they are, the more likely they are to shoot and signal to the brain that blood pressure is too high, and it is time to activate the vagus nerve to decrease the blood pressure. One research tested the effect of the slow yogic breathing known as ujjayi on seventeen stable people at different speeds of breathing in and out. To roughly equal breathing time, ujjayi respiration increases baroreceptor sensitivity and vagus activity that reduces blood pressure. This kind of slow respiration involved 6 breaths per minute, which was approximately five seconds per inhalation and five seconds per exhalation. Some scientists believe that slow yogic breathing can also reduce anxiety by reducing the sympathetic nervous system, but that is not yet confirmed.

Tip: Yoga teachers suggest you have to breathe slowly out of your abdomen. Which means your abdomen will grow or go out as you breathe in. You should cave in when you breathe out your belly. The larger your belly grows, and the more it grows, the deeper you breathe. Slow and deep breathing will induce slightly calm nervous activity. Yogis suggest you should try to take in 6 breaths a minute from your heart.

Laughter

The saying "laughter is the best medicine" may be accurate. A couple of studies suggest the health benefits of laughing. Scientists believe that laughter will activate the vagus nerve, which means that laughter therapy is good for your wellbeing. Yet experiments are still minimal, and it is difficult to say exactly how and why we feel so good with laughter. A review of yoga laughter showed that the laughing community had improved HRV (heart rate variability). There are, however, several cases of people fainting with laughter. Doctors point out that this may be too much stimulated by the vagus nerve / parasympathetic system. For example, some research suggests that after laughter, urination, cough, swallowing, or bowel movements can occur, which are all supported by vague activation. Cases of laughing people who have a rare syndrome (Angelman's), which is associated with increased vagus stimulation, are reported. Laughter is also sometimes a secondary effect of vagus nerve stimulation in children with epilepsy. Many people want to know if a good laugh is beneficial for the safety of cognitive function and heart disease. Few experiments have shown that humor improves beta-endorphins and oxides, which support the vascular system potentially. Riding can activate vagus nerves and have other health benefits, but overdoing can, in rare cases, lead to fainting.

Prayer

A small study found that reciting the rosary could lead to vague activation. In fact, cardiovascular rhythms seemed to increase, diastolic blood pressure decreased, and HRV improved. According to one research group, reading one Rosary cycle takes about 10 seconds and therefore causes people to breathe for 10 seconds (including both in and out of breath), which increases HRV and thus vagus function. Prayer slows and deepens respiration that tends to stimulate the nerve of the vagus.

PEMF

Some researchers assume that magnetic fields can stimulate the vagus nerve. Pulsed

Electromagnetic Field (PEMF) treatment enhances the amplitude of the heart rate and vagus relaxation in a sample of 30 healthy men. Nevertheless, these results have not been repeated by other research. PEMF systems are known as health products in general. The FDA has not licensed them for any reason. PEMF therapy can increase the function of vagus nerves, but more work is required.

Probiotics

Emerging evidence indicates that the gut microbiota has an effect on the brain. The intestinal nervous system connects to the brain via the vagus nerve described as "at the interface of the microbiota-gut-brain axis." Several animal studies have explored the potential effects of probiotics on the vagus nerve, but there are still no clinical trials. Animal research has shown some positive changes in the GABA receptors that were mediated by the vagus nerve of mice that had the probiotic Lactobacillus rhamnosus. GABA receptors are involved in mood in the brain, a potential link between L stimulation of the vagus nerve intestines. Rhamnose and increased GABA function add new evidence of the potential health benefits of probiotics to an emerging organism.

Exercise

Mild practice stimulates intestinal flow in animals, and vaguely activated nerves were necessary to trigger this response. Some scientists also believe that exercise will activate the vagus nerve, but there is no clinical evidence to support that.

Massage

The vagus nerve can also be stimulated by massaging certain areas such as the carotid sinus (located on your neck). Research suggests that hallucinations can be reduced. (Note: carotid sinus massage is not advised at home because of possible fainting and other hazards). The vagus nerve can also be stimulated by a pressure massage. Such massages have helped children gain weight by relaxing the intestine, and this is expected to be primarily induced by vagus nerve stimulation. Foot reflexology massages are also said to increase vagal activity and cardiac variability while decreasing cardiac speed and blood pressure, according to a small study on healthy and heart disease patients. Massages of the neck, foot, and pressure can stimulate the vagus nerve.

Fasting

Intermittent fasting and calorie reduction both increase animal heart rate variability, which is thought to be a marker of vagal tone. Others claim that irregular tempo increases their heart rate variation, but no clinical trials can support this. According to one theory, the vagus nerve can reduce metabolism by fasting. In fact, a reduction in blood glucose and a decrease in mechanical and chemical intestinal stimuli are seen in the vagus. This appears to increase the vagus impulse from liver to brain (NTS) and, according to animal data, slows down the metabolic rate. Animal studies show that hormones such as NPY rise during fasting while CCK and CRH decrease. Upon feeding, the reverse can occur. Stimulative signals from the gut-associated with satiety seem to lead to increased sympathetic function and stress reactiveness (higher CRH, CCK, and lower NPY). If they are starving, the vagus nerve may make animals more responsive to estrogen. For female rats, quickness increases the number of estrogen receptors that can be activated by the vagus nerve in certain parts of the brain (NTS and PVN). Fasting can slow down metabolism by encouraging the vagus activity of the nerves.

Sleeping or Laying on Your Right Side

Few studies suggest that lying on your right side increases the amplitude of the heart rate and the vagal response more than on other sides. Putting on the back led in one study to the lowest vagus activation. More work on this is required.

Tai Chi

Tai chi improved heart rate variation and thus possible vagus activation during a 61-person sample.

Seafood (EPA and DHA)

Omega-3 fatty acids EPA and DHA are stated to improve heart rate variability (HRV) and lower heart rate according to several research studies. HRV is directly linked to the activation of the vagus nerve. Some scientific researchers believe that vagus nerve activity may explain why omega-3 fatty acids are good for the heart, but research is required. Fish is also an important part of the lectin reduction diet.

Zinc

Zinc increased vagus stimulation in rats provided 3 days of the zinc-deficient diet. It is a very common mineral that is not enough for some people.

Acupuncture

Modern acupuncture points can stimulate the vagus nerve, particularly those in the ears, according to limited research. Not always healthy is acupuncture. In one report, a man died of a heart rate too low after vagus nerve stimulation. Be sure you work with a professional doctor and tell the doctor if you intend to visit an acupuncturist.

Eating Fiber

GLP-1 is a satiating hormone that activates ambiguous brain urges, slows down digestive movements, and makes us feel more full after meals. Fiber can be a good way of increasing, according to animal research.

EFFECTIVE DIAPHRAGMATIC EXERCISES TO GET RID OF STRESS, ANXIETY AND PANIC ATTACKS

Breathing is an involuntary body function regulated by the brain's respiratory center. When we are under stress, our respiration rate and pattern change as part of the "fight or flight response." Luckily, we have the power to change our own breathing intentionally. Scientific studies have shown that respiratory regulation can help manage pain and tension. Breath awareness is also used in yoga, tai chi, and some forms of meditation. Most people using their breath for calming and stress reduction. The primary role of respiration is oxygen intake and carbon dioxide emission through lung activity. The diaphragm (a muscle layer below the lungs) and the muscles between the ribs control the lung movement. When a human is under stress, his respiratory behavior changes. Normally, an anxious person takes small and shallow breaths and pushes air in and out of their lungs with their hands rather than their diaphragms. This breathing technique disturbs the equilibrium of gasses in the body. Shallow rapid respiration or hyperventilation may exacerbate anxiety by exacerbating the physical symptoms of stress. Breathing regulation can help improve some of these effects. You may have learned that deep breathing will ease tension. Solid research has shown that breathing techniques are not only highly effective at reducing stress in our lives, but are incredibly easy to learn and use at any time.

Breathing Exercises to Reduce Stress

How to Breathe Properly

It may sound odd, but many people don't really breathe. Natural breathing includes a large muscle in your abdomen, the diaphragm. Your belly should expand when you breathe in. Your abdomen will collapse as you breathe out. This is called diaphragmic respiration. Over time, you forget how to breathe in this way and use your chest and shoulders instead, causing short and shallow respirations that can increase your stress and fear.

Instructions

Luckily, it's never too late to learn how to relax again and how to guard against pain. Practice this simple exercise to strengthen your diaphragm: either lie on your back or sit comfortably.

- ✓ When you are sitting down, make sure you keep your back straight and relieve the anxiety by letting it go.
- ✓ Close your eyes. Close your eyes. Instead, you should keep your eyes open (and probably you can), but closing your eyes lets you focus on breathing mechanisms and not external stimulation.
- ✓ Place your stomach on one side and your shoulders on the other.
- ✓ Take a couple of breaks as normal. Is your belly rising and falling with each intake (inhalation) and every breath (exhalation)? It's perfect if you can say "yes." That's the natural way to breathe. If you rest in your uterus, but your chest rises and falls with every breath, breathe only by allowing your lungs to rise and fall while you breathe in and out.

✓ Continue to take deep breaths and focus on moving your belly only.

✓ Keep going as long as you want.

Tips

Taking these tips into account when practicing diaphragm / deep breathing: relearning how to breathe can take time.

✓ The more you do, the easier it becomes. Take some time every day to do this exercise. The nice thing is, everywhere you can do it.

✓ At a time when you are already comfortable, try to practice. This will make deeper breathing faster.

✓ If you have problems breathing deeply, try breathing in through your nose and breathing out your mouth. Slowly count to five as you breathe in and out of your mouth.

✓ In time and through practice, you will get an idea of how long deep breathing exercises are needed to reduce stress. Early on, it can be helpful to set a certain time limit, if you need time for, for example, three minutes. Keep in mind that it is usually more effective to perform multiple shorter deep breathing cycles than just long deep breathing episodes. More often, practicing often allows you to integrate deep breathing into your routine as a habit.

Other Stress Relief Techniques

✓ Once you get familiar with deep breathing, you should add more stress relief strategies in a way that works for you, including if you have a panic disorder or even fear, try to move on to 3-partite respiration to relieve serious anxiety.
✓ Talk about applying music therapy to your workouts.
✓ Consider adding images in guided images.
✓ Apply deep breathing during steady muscle-relaxing workouts.
✓ Try to breathe deeply in Pilates.

Benefits

There are many advantages of respiratory exercises recorded in research, including suppressing stress responses as they develop, allowing you to be less aggressive under stressful conditions and improving physical processes, including sleep, pain control, and even digestion. Diaphragmatic respiration is such that healthcare providers also recommend it to PTSD patients for stress reduction and emotional control. In addition, a new fitness app called BreatheWell was developed to help veterans and members of the military with PTSD and/or brain injuries recall their breathing exercises and leading them through them.

Adopting a Stress Management Lifestyle

Deep breathing is just one way to reduce stress in life or at least deal with it, but many stress management strategies are available that can help you live with more happiness and less anxiety every day. The use of a mixture of these approaches is optimal, as some strategies, in particular, are more beneficial than others. Even still, make stress management a family affair.

Breathing Exercises to Reduce Anxiety

Most people don't really know the way they breathe, but there are generally two types of breathing patterns:

- ✓ Thoracic (chest) breathing
- ✓ Diaphragmatic (abdominal) breathing

If people are anxious, they tend to breathe quickly and shallowly directly from their chests. This method of respiration is called a chest or thoracic breathing. You may not be conscious that you are breathing this way if you are nervous. Chest ventilation is responsible for a disturbance in the body's concentrations of oxygen and carbon dioxide, which results in heart rate, dizziness, muscle tension, and other sensations. The blood is not oxygenated enough, which can cause a stress response that leads to anxiety and panic. You take long, steady breaths during abdominal or diaphragmatic breathing. This is how children breathe naturally. You probably also use this breathing pattern when you're in a relaxed stage of sleep.

Difference Between Chest and Abdominal Breathing

The best way to determine your breathing pattern is to place your upper belly on one side near the tail and your lungs on the other. Note which hand raises the most when you breathe. If you breathe right, the belly should expand and contract with each respiration (and the hand will reach the maximum). Throughout tense and uncomfortable moments, when you are more likely to breathe from the lungs, it is particularly important that you know these distinctions.

Simple Abdominal Breathing Exercise for Relaxation

The next time you feel nervous, you try the simple technique of relaxation:

- ✓ Inhale your nose slowly and deeply. Relax the back. Your belly ought to swell, and your shoulders ought to grow very little.
- ✓ Exhale through your mouth slowly. When you blow the air out, gently curl your mouth, but keep your jaw straight.

You may detect a gentle "whispering" sound while you are exhaling.

✓ For a few minutes, repeat this breathing exercise.

This workout can be repeated as often as possible. You can stand up, sit down, or lie down. If you find this activity painful or think it will make you nervous or panicky, stop now. Often people with a panic disorder feel anxiety or fear during this practice. This may be due to anxiety caused by your focus on your breathing, or without some practice, you may not be able to do the practice properly. Try it again in a day or so and gradually build up the time.

Breathing exercises to reduce panic attacks

Panic attacks are unexpected, extreme outbreaks of agitation, terror, or panic. Symptoms are debilitating. The signs are both physical and emotional. Most individuals with panic attacks may have difficulty breathing, crying heavily, trembling, and heart racing. Some people will also experience chest pain and a feeling of detachment from reality during a panic attack, which is why they think they have a heart attack. Others have stated that they sound like they have a stroke. Panic attacks can be terrifying and can reach you hard.

Here are some techniques you can use to try to stop a panic attack if you have one or if you fear that:

Use deep breathing

 Deep breathing is a symptom of panic attacks and may increase anxiety, but deep breathing can minimize panic symptoms during an attack. If you can control your breathing, you are less likely to experience hyperventilation, which can exacerbate other symptoms and the panic attack. Take your breath deep into and out through your mouth, feel the air slowly filling your chest and stomach, and then leaving it again slowly. Breathe a count of four, keep a second, and then breathe a count of four.

Recognize that you have a panic attack

 In knowing that you have a panic attack instead of a heart attack, you will note that this is acute, it's going to happen, and you're OK. Taking away the fear that you may die, or that there is an impending disaster, all signs of a panic attack. You should rely on other strategies to reduce the symptoms.

Close your eyes

Many panic attacks are causing you. If you are in a fast-paced situation, it can fuel your panic attack. Close your eyes during your panic attack to eliminate stimulation. This can block additional stimulation to make your breathing easier to concentrate.

Practice mindfulness

Practice mindfulness Treatment will help you stay rooted in the reality of what is around you. Panic attacks can lead to a sense of detachment or detachment from reality, which can counteract the panic attack as it progresses or actually happens. Focus on your physical sensations, such as digging your legs into the ground or feeling the texture of your jeans. These particular sensations are firmly rooted in reality and give you something to focus on.

Find a focus object

Many people find it helpful to find a single target in a panic attack to concentrate all their energy on. Select an object in clear sight and consciously note it all. You can, for example, note how the clock hand just when it ticks, and that it's slightly bumpy. Describe for yourself the shapes, color, types, and size of the piece. Focus on this point, all your attention and your panic symptoms will subside.

Use muscle relaxation techniques

Much like deep breathing, muscle relaxing will avoid your panic attack by regulating the reaction of your body. Consciously relax one muscle at a time and start with something simple like your fingers and move through your body. The methods of muscle relaxation are most effective when you have learned them beforehand.

Picture your happy place

What can you think of as the most enjoyable place in the world? A sunny beach with calm waves? A mountain cabin?? Imagine yourself and try to focus as much as possible on the specifics. Imagine digging your toes in the warm sand or smelling pine trees ' sharp scent. This should not be a quiet, calm, and relaxing street in New York or Hong Kong, regardless of how much you really love the cities.

Engage in light exercise

Endorphins hold the blood pumping just as soon as possible. It can help our bodies flood with endorphins that can improve our mood. Because you're stressed, choose light exercise, like walking or swimming, which is gentle on the body. The exception is whether you hyperventilate or have difficulty breathing. Do what it takes first to catch your breath.

Keep lavender on hand

Lavender is known for being calming and stress-relieving. It can help to calm your muscles. When you think you are going to be anxious, take some essential oil from lavender and put it on your forearms when you are stressed. Breathe in the fragrance. You can also drink chamomile or lavender tea. We are both relaxing and calming. Benzodiazepines should not be mixed with lavender. The mixture will cause intense somnolence.

Repeat a mantra internally

It can be relaxing and reassuring to repeat a mantra internally and give something to grasp in a panic attack. Whether it is just a "this too" or a mantra that appeals to you, repeat it on the loop until you know that the panic attack has begun to subside.

Take benzodiazepines

Benzodiazepines can help treat panic attacks if you take one as soon as you feel that there is an assault on them. Although other approaches to the treatment of panic may be preferential, psychiatry has recognized that there are a few people who will not fully (or in some cases at all) respond to the other approaches above and, as such, depend on pharmacology treatment approaches. Such methods often include benzodiazepines, some of which are approved by the FDA to treat this disorder, for example, alprazolam (Xanax). Since benzodiazepines are a drug given, you would actually need to treat a panic disorder in order to take the prescription. This drug can be extremely addictive, and over time the body can adapt to it. It should be used only sparingly and in extreme situations.

WAYS TRAUMA CAN AFFECT THE NERVOUS SYSTEM AS WELL AS PREVENTION TIPS

Brain and body have evolved after all forms of damage (from war to car accidents, natural disasters and domestic violence, sexual assault, and child abuse). Each neuron retains memories, and every trauma-related, embedded neuropathy has the opportunity to reactivate again and again. The shifts such imprints produce are sometimes transitory, the small glimpse of unsettling visions and moods that subside within a few weeks. In other cases, changes are readily apparent signs that conflict with jobs, interactions, and relationships and impair their operation. One of the hardest things for patients after trauma is to understand the changes that come, and to incorporate their meaning, how they affect their lives and what can be done to improve them. The recovery process starts with the normalization of post-trauma symptoms by analyzing how the damage affects the brain and the complications caused by these effects. The reptile brain takes charge, moving the body to defensive mode during a traumatic experience. The brain stem, which shuts down all non-essential body and mind functions, orchestrates survival mode. During this time, the nervous system raises stress hormones and prepares the body for battle, escaping, or freezing. In a normal situation, the parasympathetic nervous system switches the body into a restaurant state when the immediate threat stops. This technique reduces stress hormones and allows the brain to revert to the normal, top-down control structure.

Nevertheless, the change from the reactive to the receptive state rarely happens to these 20 percent of trauma survivors who develop symptoms of posttraumatic stress disorder (PTSD)— an unmixed experience of anxiety associated with past trauma. The reptilian brain, which is predominantly endangered and protected by dysregulated behavior in major brain systems, retains a lifelong reactive survivor.

Everybody's traumatic events happen. So sometimes we get traumatized even when it doesn't happen to us personally–you learn of another high school shooting, a drunk driving crash, or an attacker killing people. You will continuously see it on social media: allowing other people to engage with our emotions makes us feel less alone in pain–whether it happened to us, we saw it, or we saw it on the television. Often our brain doesn't understand the difference. As people, we generally treat pain in two ways. The condition of hyperacoustics–your brain runs, you can not fall asleep or relax, and the accident can begin again even after it has happened, which can interrupt your ability to concentrate. The hyperarousal condition is on the flip side— you feel numb and lethargic if you look at it. The nervous system is adaptive and stress-controlled at all stages. We all have a "battle, flight or freeze" reaction without our voluntary knowledge or influence. Actually, we often don't know why it happens or why we react the way we do it. Sometimes we build on past traumatic events unintentionally we may even have overlooked. If I deal with a stressful client, I sometimes find that their challenge, run, or freeze responses are somewhat fussy. It is like the warning mode of your nervous system is for a predator–i.e., they are about to be targeted. Such consumers share their feeling that they are hypersensitive to their environment. This often causes them to become unnecessarily defensive, arguing, or verbally militant–or to be shut up, disappear, hide or flee social situations. Unlike humans, we do the same when we experience emotional or physical threat as the gazelle. It is possible to switch between cooperative grassing, combat, or flying (sympathetic machine battle and flight) or shut-down (parasympathetic-shut-down mode). Our answer lies in our interpretation of the case. Perhaps someone just played a game when they jumped out to frighten us, so we fainted. No matter if the accident was intentional or not, our body moved to stop, we saw it as a wound. Our body moved to shutdown mode. Or perhaps the stress incident was really life-threatening, and the sensations were properly responded by our nervous system. Regardless of the cause, our brain thought what happened was life-threatening enough to make our body fight, run, and shutdown. If someone has undergone a traumatic event that leads to a breakdown, any circumstance that reminds people of a life-threatening situation will cause them to be detached or disassociated again. Individuals can even survive for days or months in a state of disconnection or shutdown. This is often felt by veterans during loud, sudden noises like fireworks or thunderstorms. A raped woman may easily become hyper-vigilant or dissociated if she thinks anyone watches her. Anyone hurt could be shocked when even someone else cries. The problem happens when the initial trauma has not been handled in such a manner that the original trauma is fixed. This is what PTSD (post-traumatic stress disorder) is — the overreaction of our body to a small answer and either trapped or shut down in combat. Those with traumas and shutdowns usually feel guilty for their inability to act because their bodies have not changed. They would often have preferred to have fought more during that period. A Vietnam veterinarian may feel that his companions who died around him, frozen in fear, failed. A rape victim may believe that they haven't resisted their attacker because they have frozen. A victim of abuse may believe that she is trying to escape the attacker and that she is vulnerable or ineffective. Much of' pain' preparation, which teaches people to remain in battle and flight mode, is intended to prevent people from being separated during actual life or death scenarios. Such activities are sadly not popular among professional sports teams or special forces. The right amount of stress will drive our nervous systems to greater tolerance with a good recovery.

How Sexual Trauma Relates To Polyvagal Theory

The polyvagal theory stresses that the way we react to the world depends on our physiological condition. This is important for people experiencing sexual trauma. This is important. When sexual trauma survivors stop or withdraw, their medical state changes. The adaptive nervous system changes the way it controls the body's organs. In this modified condition, the world outlook of the victim is very partial. The trauma-induced change in the physiological state leads to trauma patients viewing nearly everyone as a threat from the polyvagal viewpoint. The psychological history of sexual assault survivors also suggests that they want marriages, but they find it difficult to accept and be close. Their bodies will not tolerate closeness and comfortable physical contact. The polyvagal theory explains the bio-behavioral, neurological, and traumatic experiences. The polyvagal theory also gives clues to counteract these weakening features. This is achieved by working on techniques to change the physiological status to make the individual relaxed and healthy.

Trauma recovery and the vagus nerve

It is important to understand what occurs after a wound on a neurological level. The vagus nerve plays a major role in helping people recover from injuries and restore composure. The vagus nerve is an important part of the parasympathetic nervous system, which calms the organs after the initial risk of adrenaline. The vagus nerve plays a crucial role in the overall health service. It tells the brain what happens in our bodies, particularly the digestive tract (stomach and gut), lung and heart, spleen, liver, and kidney. Some of us have stronger vagus activity, which means that after stress, our bodies can return to normal faster. On a neurological level, the first reaction of our body comes in the subcortical areas of the brain, and only then do we use our conscious mind in the higher cortical regions to perceive those stimuli, to make conscious choices. Trauma may be weak, but one of the most important things I try to teach my customers is that even long after the trauma has taken place, they may do something about it. Although their initial trauma response was automatic, they can now decide to make a conscious choice. It is appropriate to distinguish people from their capacity to choose various actions following trauma. However, by using your parasympathetic nervous system, you can help reorient your brain and change your vagal tone. I often find it difficult not to respond sympathetically (stress) with high stress but to inhibit its parasympathetic (rest and repair) system. If your rest and reparation mechanism doesn't work properly, it can affect everything from appetite and metabolism to energy level and sleep. This is important because sleep is usually the dominant feature of your rest and repair system—so that your body can recover and heal.

So how do we get out of shutdown? The social engagement system is the reverse of the dorsal vagal system. Simply put, what solves shutdown mode leads to a healthy social relationship or an acceptable connection. Looking inside our bolts and nuts will help us understand why we feel the way we do when the body is battling, running, or shutting down. If we understand why our body reacts in the way it does, we can understand how to change states like a string of clues and basic brain science. We should start moving out of battle or travel, out of suspension, and into the state of social engagement. Whether we are only making a connection with a new anxiety patient or helping them to deal with their deepest traumatic memories, it is important to know how to navigate the polyvagal states. It may also be helpful if some of these symptoms have just been identified by you. Of starters, "I feel lightheaded and detached when I am with my parents, even when an adult and they are starting to fight." If you have seen these symptoms in yourself, ideally through counseling and even learning how it works, you may be able to pull yourself away from an unconnected condition. Research has shown that some parts of the brain are shut down during stressful experiences recovery, including speech centers and cognitive thought centers. It is, therefore, necessary to practice therapy in a stable, healthy, and safe environment or to exit the shutdown phase. Therefore, it is important to have a strong attachment. Otherwise, the patient is at risk of trauma. Since I am a psychiatrist, I am writing this to show how a patient can exit the shutdown mode.

Such guidelines are, however, only applicable to people who understand how shutdown mode operates. And it can even benefit people who feel shut to learn how to try again to reach a balanced mode of social engagement.

Have a trust-based relationship. Due to the ability to re-traumatize, don't even deal with traumatic events— especially if you feel the shutdown mode began until the therapy process is deeply connected. As a therapist, it is important to allow a patient to express things that they can't communicate to others— sorry thoughts, rage, emotional response, anything that feels uncomfortable to share with others.

Find your own calm center. You give them a lifeline if you can sense their distress, sit with them, and help them to feel connected through their shutdown. You help them get out of the freeze, through social activity. It is necessary to fight the desire to dissociate, however terrible the subject is. As clinicians, we will detach ourselves because of the synaptic reaction to a mirror — to replicate our patient's head, and because it's easy to imagine it occurring to us when we experience horrible trauma. The human experience is so strong that, as we take the pain back in order to help us, it re-writes it in our subconscious and incorporates the sense of being helped in the recollection of the trauma. We are forming new neural pathways around the trauma and can alter the reaction of our body to it.

Let the patient lead. Don't go looking for a witch. Step into the issue if the patient brings it up. Yet, it is dangerous to direct the patient to something that does not happen by questioning and attempting to get him to confess. Don't let your own experience drive you to think you have anything too.

Normalize their response. The whole idea of polyvagus will make us say, "thank you!"To our souls. To our heads. Even if this system is sometimes overactive— unfounded panic or anxiety— our body watches over us and tries to keep us alive. The body reacts in this manner, as the gazelle either runs away or limps. And gazelles don't know what emotions are first and foremost. Now that the individual knows that their emotional response has been pragmatic, dominant, and necessary, we can get rid of the embarrassment created by their non-reaction.

Help them find their anger. Anger is an unbelievably resilient trait, and we don't allow it to be. We agree that rage is evil. But really, anger shows us where our healthy limits have been crossed. Wrath gives us the strength to overcome this challenge. We can make a patient understand that he has the emotional energy to resolve, but at the time he needed it, the energy couldn't manifest. If we can get a participant to recognize their frustration in a session, they will see they did not respond fully to the traumatic event. If we can help them feel the slightest movement of a micro-expression of anger on their faces— a slight decrease of the inner eyes— that's when their bodies didn't completely deceive them. We can reconnect your body and your feelings to your emotions. It helps to develop a state of congruence— where your inner feelings complement your outward feelings.

Often, rage and guilt management helps the mind to fundamentally change as a dissociative experience is discussed. Anger drives them out of dissociation, even though the doctor is upset at you!

Introduce body movement. Due to the fact that pause allows us to freeze, it is a great way of reconnecting mind and body to reactivate body motion while thinking of trauma. It is important to move attentively and gradually, focusing on the sensation of movement. The patient's movements–slow hitting, jumping, spinning, slow running–break the individual from the shutdown into fight or flight mode, with the goal of bonding or social engagement. Body movement techniques can fundamentally change memories in tandem with listening to a therapist.

Practicing assertiveness. Emotional shutdown in relationships may occur when one person feels that he or she can not connect well with another person. This practice is described as stonewalling by a therapist. Practicing assertiveness can help the patient to regulate their emotional state and to step comfortably into healthier forms of relationships.

Breathwork, mindfulness, and yoga, both play a role in getting closer to your body here and now. Teaching yourself how to defend yourself better in the future can be effective and, over time, resets the tension mechanism. More doing something intense, constantly, allows you to build an inner strength that can keep you going and fighting longer until you shut down.

Become a Judo Master and practice strength training. Teaching how to protect yourself better in the future can be powerful and also over time, reinstates the stress system. Anything challenging on an ongoing basis allows you to build an inner strength that can sustain you struggling and flying longer until you rest.

How MELT works to help heal trauma

The cornerstone of MELT is the rebalancing process that stimulates and allows the parasympathetic nervous system to prevail while you are alive. That is why the Rebalance Process impacts all facets of your body in such a broadway. The equilibrium series is typically one of the first therapies I suggest. A key technique in the 3D Breath Breakdown sequence is inhalation. Inhalation. While breathing is uncontrolled and unconscious, we can actually control it consciously. We deliberately participate in the sympathetic response. By slowing down and concentrating on the direction of diaphragmatic motion, you change how the brain stem signals the diaphragm to contract as you respire and don't think about it about 25,000 times a day. Instead, we add another method called 3D Breath, which stimulates the central neuronal representation in an exhalation. It increases parasympathy, improves overall pain regulation, and stimulates vagal tone. Through when you breathe in your heart, the supply of oxygenated blood through your body is speeded up. Breathe out, and the pace slows down. Cardiac variability is one of many things regulated by the vagus nerve that works when you respire but is suppressed when you inhale, so the greater your heart rate difference, the higher your vagal tone. If you are recovering from depression, follow the Harmony Series, and see if you can change your thoughts in just 10 minutes. The sequence is so easy and efficient. I recommend performing it every day

at the beginning or at the end of your day for
the best results.

SUREFIRE WAYS TO PRACTICE THE POLYVAGAL THEORY IN YOUR DAILY LIFE

If you do not have an external surgical unit, you can not activate the vagus nerve directly; however, it can partially trigger the vagus nerve to release keyed up or nervous system conditions. Note, it moves through your bowels, diaphragms, lungs, mouth, inner ears, and facial muscles. Practices that alter or regulate the functions of these areas of the body can, therefore, affect the activity of the vagus nerve through the internal body feedback loop.

Humming: The vagus nerve travels through the vocal cords, the inner ear and the sounds of the music are a free and easy way to influence the state of your nervous system, you should check this out in your living room. Just choose your dream song, and you're good to go. Or you can "OM" the path to wellness if the yoga suits your lifestyle. Notice and enjoy your chest, throat, and head sensations.

Conscious Breathing: breathing is one of the quickest ways of manipulating our nervous system symptoms. The goal is to slow the belly and diaphragm down with the air. Vagus nerve stimulation happens as our normal 10-14 breathes slow down to 5-7 breaths a minute. You can do this by measuring the inhalation to 5, retaining it quickly, and extending to 10. You will activate the vagus nerve further by causing a small constriction behind the throat and producing a "hh" Try to fog a mirror in order to create the feeling in the throat and inhale it out of the sound of the nose (in yoga this is called Ujjayi pranayama).

Valsalva Maneuver: This complex name refers to the exhale of a blocked airway. This you can do by keeping your mouth closed and blowing your nose. This increases the pressure inside the chest cavity and increases the vagus tone.

Diving Reflex: Considered to be a first-rate nervous stimulation technique, the cold water that springs from your lips to your scalp stimulates the reflex of diving. You can also achieve cooling effects for the nervous system by placing ice cubes in a ziplock and holding the ice against your face and breathing. The diving meditation slows your heart rate, raises your brain blood flow, decreases frustration, and relaxes your body. One strategy that activates the reflex of swimming is to dip the tongue in air. Drink and keep tidal water in your mouth, feeling your tongue with the water.

Connection: Relationship access. Whether it happens in person, over the telephone, or even via text or social media in our modern world, healthy connections to others can trigger the regulation of your body and mind. Relationships can invoke the spirit of fun and imagination or can ease us into a comfortable friendship with others. You might have a lighthearted exchange of messages with a friend. If you're around someone else, you should test your relationship expert, David Snarch's simple yet powerful technique, which is called "hugging before calming." The guidelines are clearly to "stand up on your own two feet, put your arms around your friend, concentrate on yourself, and keep quiet and relaxed.

Extend the Duration of Your Exhale: " Use the breath as an anchor when you feel stressed or overwhelmed and concentrate on it. Slow exhalation encourages the supremacy of the parasympathetic system, which can help to return the body to a calm state of mind. The increased vagal force slows our pulse, the pace that has a soothing effect if we prolong exhalation. When we are nervous or panic-stressed, our typical response is to take short, shallow breaths that raise the heart rate by reducing the vagus effect and make you feel less grounded and anxious. Practices such as meditation and pranayama yoga cause vagotropic effects, which slow down breathing and pulse rate and increase the blood flow to the heart. Evoking relief by mental-body rituals such as transcendental meditation and prayer are perfect ways to reduce the release of carbon dioxide and slow down the body: stretching your exhalation for 2-4 seconds longer than usual is enough to have an impact. Delete your breath to let your body feel fear and stress.

Stand Up and Bring Awareness to Your Posture: "stand up switches our alertness and causes blood pressure changes," says Dr. Porges. In many classes of yoga and meditation, you might wonder why body alignment and posture are highlighted. "Earth" is a common term in mind-body activities, which means the diaphragms can be completely exhaled and comfortable. When you sit in front of a computer screen or walk to a car for long periods of the day, then it may be time for you to take a serious look at how you behave. It can change your awareness and energy simply by adjusting your posture. Through maintaining good posture during long exhalation, the parasympathetic nervous system is activated. Although you might believe it is oh so easy to just let gravity do its work, it requires consistent practice to maintain an isometric posture that will strengthen your core and allow you to relax during any activity.

Really Listen to Your Body: you might be intrigued to see what the founder of the Polyvagal Theory does to encourage a distinctly sound every day. His response in effect was very simple: "What I do is that I am clinically aware of my body, so I get very much aware of changes in my body and try to respect these shifts and move my body physically to conditions where it will feel safe. Sometimes you're immediately interrupted, and if you have sufficient control of the climate, you basically say,' Mind, I am going to move to a quiet spot.' But you really don't have this option when you're in certain work environments. What I'm doing now is assess what my body can do and, if things are stressful, I'm trying to be sure that I have a rest, a relaxed moment when I'm in a quiet place. In our go-go culture, which is infatuated with instant gratification and glory, it is vital for health to ensure continuity between body and mind. It is now palpable more than ever that the prescription for a life-long living is to strike a balance between the many competing priorities of life and to take into account the change in your body and your physiological condition that is more effective than we have always imagined for you (and others!).

Sing and Listen to Music with Other People : A central principle of Polyvagal Theory is that we are connected, as the specific (i.e., ventralized) vagal pathway that originates in an area of the brainstem that regulates striated face and head muscles, involved in hearing, vocalization and facial expression, is firmly rooted in our relaxed state. This system allows us and other mammals to be social, according to Dr. Porges. "Song performing and listening are other outlets for improving the culture," said Dr. Porges. If you jammed into a favorite song while driving on an open road, you know how fast your mood can change with music and singing. In addition, Tibetan lamas maintain high amounts of carbon dioxide during chanting expansion. Sociality plays a major role in ensuring the safety of our nervous system from damage. Porges coined the term "neuroception" to describe the way in which our neural pathways identify circumstances and persons of health and threat without conscious awareness. Security neuroception is important before social engagement patterns can grow for social bonds. Perception is a body consciousness and not a perceptual understanding of the signals that are driving our physiology into a different phase,' says Porges,' that occurs at a point beneath our consciousness.

HOW TO STIMULATE THE VAGUS NERVE WITH NATURAL TECHNIQUES

The longest cranial nerve, which encircles back, thorax, and belly, is vagus nerve (VN) with afferent fibers that relay a spectrum for somatic structures and independent preganglionic of interoceptive inputs and efferent fibers. Electrical stimulation of the VN using implanted systems (VNS) has been established in recent decades, contributing to its acceptance for epilepsy and depression care. Recently, non-invasive techniques were developed for VN stimulation. The NV has many functions, and the operation most conducive for measurement is its effect on cardiac rhythm regulation. The assessment of heart rate variability (HRV) can be quickly measured. Higher vaginal parasympathy results in the decline of HRV and is associated with a wide range of conditions of health, including a higher risk of early mortality. Epileptic patients, particularly those with poorly controlled seizures, were shown to have parasympathetic tone impaired. Could natural ways raise VN, as shown by improved HRV, enhance parasympathetic speech, strengthen seizure control? There are many natural ways of enhancing the VN, raising the HRV, and, therefore, the part sympathy tone. These are normal ways of reducing stress, fitness, and sleep, primarily in 3 groups. Although the natural ways of stimulating VN have shown an increase in HRV, seizures have not been reduced. The exception is Mozart's composition, which has proven to increase parasympathy and decrease convulsions. There is a strong need for much more research to analyze the effects of the various methods of that HRV on seizures.

Thankfully, we can do some things by ourselves to optimize the connection between the brain and the intestine via the vagus nerve. The following steps will help to control vagal toning, reduce inflammation (which can inhibit distinctly nervous function) and guarantee a healthy overall combination between parasympathy and compassion. It, in effect, will help you recover better from intense cycles, improve your metabolism, and contribute to a host of other full-body benefits.

Try deep breathing or meditation (or both!).

Deep respiration is one of the simplest yet most effective ways of activating the vagus nerve. If your exhalation is even many times longer than your inhalation, the slight neuron sends a signal to your brain that your parasympathetic nervous system is turned on. Consider this: Inhale for 2 minutes, then four counts, with an inhalation break at the top and an inhalation break at the bottom of the exhalation. Multiple studies also confirm mediation power to increase pain, sleep, anxiety, and gastrointestinal function through a direct effect on vagal tone.

Head to a yoga class.

Research demonstrates that frequently practicing gradually, such as yoga, improves gastric motility, contractions of the smooth gastric muscle necessary for food movements through the digestive tract, and this is achieved by activating the vagus nerve.

Take a cold shower.

Take a one minute blast of cold water to finish your shower and do not be afraid to go out for a brisk stroll. Studies show that the vagus nerve and separate neurons in the vagus nerve pathway are triggered by intense cold exposure, inducing a change in function in the parasympathetic nervous system.

Eat foods rich in tryptophan.

Dietary tryptophan is metabolized in the gut and can help the astrocytes — brain and spinal cord cells — manage inflammation that can improve communication via the vagal messenger path from the gut to the brain. Spinach, beans, nuts, bananas, and poultry are included in the meal.

Maintain a healthy weight.

Obesity and intestinal inflammation may interrupt vagal function and negatively affect the brain-GI relation. If you are overweight, then your best bet is to adopt a sustainable practice that will result in long-term loss of weight. My advice: move your body every day and concentrate on a diet rich in various fruits and vegetables, nuts, seeds, and legumes, including the Mediterranean diet.

Make sure you poop daily.

Consume plenty of fiber-rich foods every day (targeted for more than 25 grams) and maintain your normal sleep and exercise habits in order to eliminate your body at every rhythm. Healthy removal means the toxic food waste in the colony is less stagnated, and that undesirable species have a less hospitable environment, which may impede contact between the brain and the gut.

Nix sugar from your diet.

Excessive sugar causes not only chronic inflammation but also affects synaptic feedback, and other signaling pathways and pathogens may continue to perpetuate inflammatory signals in the brain through inflammation in the GI tract.

Pop a probiotic.

In addition to cutting sugar consumption, consider adding fermented food or probiotics in your diet to promote a healthy intestine, and sustain optimum intestinal brain signals. Work has shown that intestinal microorganisms can activate the vagus nerve. One study showed increased GABA production and a reduction in stress, depression, and anxiety of mice who were given the probiotic Lactobacillus rhamnosus. This beneficial effect, though, was not seen in mice whose vagus nerves were cut.

If you eat loads of animal protein, scale back.

Red and choline-like meat and eggs are converted into trimethylamine N-oxide (TMAO) when consumed in excess, a substance that is linked to inflammation and cardiovascular disorders. Reduced consumption of these foods can reduce inflammation and better regulate parasympathetic and sympathetic vitals such as blood and heart rate.

Consider intermittent fasting.

In some research, fasting and dietary restriction can stimulate the vagus nerve. Yet despite the many other benefits of fasting — from increased cognitive function to weight loss yet decreased inflammation — it could be worth trying. The best thing: The fasting time doesn't have to be too long to benefit greatly.

Belt out your favorite tune.

Research shows that singing has a calming biological effect, which is all about the vagus nerve. Go on, turn on the radio when you're in the car – or even better, if you take a cold shower!

THE UNEXPECTED ROLE OF THE VAGUS NERVE IN CHILDREN'S DEVELOPMENT

A relationship between vagal behavior and child growth has been reported by several studies. Baseline vagal activity in premature and full-term infants is related positively to age parallel to normal autonomic nervous system maturation. Similarly, the level of maturation and autonomy of the autonomic nervous system is related to child vagal activity. E.g., premature infants have high basic vagal activity than full-term children, and infants with lower basic vagal activity appear to have less favorable neurodevelopmental outcomes.

Vagal cardiac control is used as a proxy for risk babies, including very low birth weight preterm infants. In a study on low birthweight babies, vagal activity and maturations in vagal activity between the ages of 33 and 35 weeks are observed. These distinctly active measures expected outcomes above birthweight, disease complications, and socioeconomic status over three years. Higher vagal activities were related to better social skills, while higher vagal growth was linked to increased mental processing and gross motor skills. When the birth weights of the sample were lower than 1,000 grams, compared to those of birth weights of more than 1,000 grams, the vaguely mature activity became a strong predictor in less than a thousand grams of mental processing, knowledge base, and gross motor skills. Neonatal risk factors were not linked to school-aging outcome measures in six to nine years of follow-up to a sub-sample of this very small group of birthweight infant children, while vaginal maturation coincided with social competence, as measured in the Child Behavior Checklist. A similar examination of very early vaginal development with low-risk fetuses was also carried out between 36 and 40 weeks of gestation. The authors used respiratory sinus arrhythmias to measure vagal activities and quantified the efficiency of homeostatic control for each child with a Slope (SRSA) and RRSA regression line correlation coefficient, which related to cardiac fluctuations and RSA fluctuations. They tested the association between RSA and both SRSA and RRSA in low-risk fetuses for their hypothesis. We observed that the fetuses controlled by parasympathy had higher SRSA and RRSA values and were more powerful homeostasis regulators than supporters.

Vagal Stimulation for Preterm Infants

Infants may be activated by vagal stimulation and development, and relaxation such as kangaroo care and massage therapy may be noninvasive ways to improve baseline vagal activities in babies. A recent study on kangaroo treatment, for example, increased the ripening of vagal activity. In the research, skin-to-skin (kangaroo) impacts on mobility, state regulations, and neurobehavioral status of preterm infants providing kangaroo treatment for a span of 24 days have been investigated. The baseline vagal movement was determined from the heart rate ten minutes before kangaroo care started, and from the gestational age 37 weeks again. Children who provided kangaroo care displayed a more accelerated maturation from 32 and 37 weeks of gestational age, as well as stronger state institutions with long sleep and alertness and shorter productive sleep times. The results on the Brazelton Neonatal

Behavior Assessment Scale habits and attitude components have indicated a more advanced neurodevelentation profile.

In Korea, a group measured responses to stimuli, including vaginal baseline operation, heart rate, and oxygen saturation with a different form of skin contact, that is, infant massage (tactile and kinesthetic stimulation). Past babies have been randomly assigned to massage for a standard treatment control group twice a day for a span of ten days. After the massage, vagal activity was significantly higher than in the experimental group before the massage, while there was no improvement in function. The therapy group often took considerably longer to be awake and active. Therefore, baseline vagal activity was associated with more attention and organized behavior.

A significant weight gain has been identified in many studies on massage therapy for preterm infants. The massaged premature neonates did not take more calories than the control neonates in these studies, nor did they preserve more calories by resting. Since excluding those possible underlying mechanisms, we explored the probability of increasing vaginal movement after massage therapy leading to increased gastric motility and thus contributing to weight gain. In this study, the vagal movement was assessed in relation to massage (a high pressurized massage group) with gastric motility versus placebo massage (a low-pressure massage group) and a control group for a 5-day time period. Premature neonates receiving moderate pressure massage therapy showed a greater increase in weight. There was regularly improved vagal tone and greater gastric motility for these children during and immediately following the therapy sessions. Vagal tone and gastric motility were significantly related to weight gain during massage therapy. In the other previous studies reviewed by Field et al., the weight gain undergone by preterm neonates could, therefore, have been caused by decreased vagal movement and gastric motility. In addition to improved gastric motility, an improvement in vagal activity could also lead to an increase in food intake hormones, which would also help to increase weight gain. Food intake hormones (i.e., insulin) and digestive hormones (i.e., gastrin) are activated by vagal stimulation.

Therefore, the release of both food absorption (insulin) and digestive hormones is facilitated through tactile stimulation.

Infant Affect and Early Interactions.

Emotional signals were associated with autonomous action, but the autonomous activity was measured alongside facial expressions only over the last decades by many famous scientists, including Charles Darwin (1872), Williams James (1884), and Walter Cannon (1927). In research in the 1980s, for example, newborns with more vocal heart rate variability have also been observed, and three months-olds with a higher heart rate variability have shown more concern. Older 5 months old children with more vagal activity showed more interest and joy, while children with lower vagal tone showed more look-alike behavior.

In the sense of mother-to-child relationships, vagal behavior was also measured. In a study, child facial expressions were coded using the facial expression coding system of AFFEX, and EKG of infants was recorded at three months of age during their interactions with mothers. Vagus behavior was associated with signs of happiness and excitement, and behavior. In comparison, the low vagal behavior in children of depressive mothers was related to sad and wrathful facial expressions. In a similar study by our party, more vocalization was associated with the higher vagal activity.

In another study, there were favorable associations between the normal vagal behavior of babies and the symmetrical patterns of mother-infant dyad encounters. EKG was obtained from the babies in the Porter study accompanied by a face-to-face mother-infant contact lasting 15 minutes. The interactions have been videotaped and coded for symmetrical patterns and disturbing coregulation patterns. In another study, distinctly newborn development anticipated the synchronous three-month-old mother-child relationship. In this study, the newborns were tested for child orientation using the Brazelton Neonatal Behavior Assessment Scale and sleep-wake cyclicity and vagus. Throughout three months, mother-infant synchrony was tested by videotape micro analyzes and face-to-face touch tests. Results indicated that in a newborn phase, sleep-wake cyclicity, vagal activity, orientation, and arousal control were each a distinct predictor for mother-infant synchrony. Vagal behavior was one of many predictor variables in this context. Disturbances in mother-to-child experiences such as the still look mother have been found to influence vagal activity. The affective states of babies and vagal behavior were reported in a report on still-face relationships between young children aged 3months and their mothers. The authors evaluated the synchrony and the correlation between the affective states of mothers and infants and monitored infant heart rate and vagal activity. The children displayed increased negative effects during the shutdown and reduced vagal activity.

Infant Temperament

Infant temperament has also been related to vagal activity. Temperament may be perceived to be the infant's stable affective mood. In an early study, we explored the threshold for tactile stimulation (pinprick), their reactions to model facial expressions (feeling, sad, surprised), and how precise the models ' expressions could be measured in their so-called "imitative expressions." We also measured the habitude of the model faces and the vagal activity. We observed the greater variation in heart rate for infants we called "externalizers" (infants with highly expressive facial expressions). In comparison, the infants we called "internalizers" (infants with flat affection) had a slight variability in heart rate. Vagal behavior could, in this way, be seen as a stable physiological state while expressiveness is seen as a stable affective state. The association between vagal activity and an early childhood personality may not be unexpected, as both disposition and vagal behavior are observed for consistency and reliability over time. In a retrospective longitudinal study, vagal development was measured in children and their mothers for 2 months and 5 years. The authors confirmed that the increase in vagal behavior in both children and mothers was continuing. The girls also had 5 years of adult vagal activity and did not differ from their mothers in the basic level of vagal activity. Baseline vagal activity appeared to be normal, but baseline vagal activities in children were unpredictable, while both mothers were stable. Similar retrospective research, but with high-risk children (low birth weight), associated the slightly developed behavior with social skills, as measured by the Child Behavior Checklist. By contrast, neonatal risk measures (low birth weight, low socioeconomic status, high medical risk) did not correlate with any of the measures in school age.

Low Baseline Vagal Tone in Infants of Prenatally Depressed, Anxious and Angry Mothers

High vagal behavior for stressed mothers and their children has been recorded as early as the neonatal stage. In a study in 1986, neonates with depressive mothers had both reduced vagal output as well as higher right frontal EEG (a proxy for negative effects and commonly reported in adults with depressed conditions). In a 2004 study suffering mothers showed lower vagal tone and a higher relative right frontal EEG activity than non-depressed mothers. Likewise, their newborns had lower vagal tones and higher relative right-front EEG stimulation. Some newborns of stressed mothers will mimic the vagal tone and EEG of mothers with their identical biochemical profiles. There were high concentrations of maternal cortisol in mothers and less dopamine and serotonin than in non-depressed women. Similarly, the cortisol levels of newborns were elevated, and their levels of dopamine and serotonin lower than those of non-depressed mothers. In the same research, a major association between lower vagus activity and higher cortisol and lower levels of serotonin and dopamine was found. The predictor variable for risk factors for prenatal depression, the variable for outcomes expected by maternal vagal activity and the associated biochemical variables, including cortisol, serotonin, and dopamine, in this case, were infant vagal activity. The lower vagal activity of the neonates may also explain their lower expressiveness, as the vagus nerve internalizes the facial nerve and then insides the face. In this case, vagus behavior could be viewed as causal or as an intrinsic communication function. We documented their faces during Brazelton's Neonatal Behavior Assessment Scale and during the simulation of joyful, sad, and shocked face in a study of infant facial expressions of anxious mothers. The newborns of distressed mothers showed lower performance in the Brazelton orientation cluster (less concentration and less recreation during the Brazelton period). During the facial expression modeling, the modeled glad and surprised facial expressions showed less orientation and fewer facial expressions. Porges has described' the intelligent vagus ' activity as a social engagement system controlling the looking at (the ventral vagal complex's somatomotor components, cranial nerve XI in this case) listening (the cranial nerve VII), vocalization (crane nerve IX, X), and facial expressions. High vagal behavior, as in depressive behavior, can be characterized by smooth facial and vocalizations without intonational contour.

Neonatal predisposition to lower baseline vaginal activity can persist as we have found lower vagal activity in the three-and six-month-old babies of depressed mothers, but a significant increase in baseline vagal activity in infants of non-depressed mothers has occurred from three to six months, and no increase for babies of depressed mothers has occurred. In this study, higher vagal activity (e.g., cooing and babbling) was associated with more vocalizing. Those results plus evidence described earlier revealed that three-month-old babies of depressive mothers had significantly greater sad and angry and fewer expressions of interest than babies with non-depressed mothers, indicating that maternal depression affects both the emotional expression and their distinctly active behavior of children. The neonates had not only lower vagal function but also greater relative right frontal EEG response, higher cortisol, and lower dopamine and serotonin, as was observed in our multivariable sample of newborns with distressed mums. Multivariable profile tests can distinguish individuals and their developmental courses more reliably than a single variable, for example, vagal behavior. The combination of these measures in infant and autism studies would not only help us to understand more about infant and individual disorders but also to understand the interactions of these complex systems in a theory of poly systems.

CONCLUSION

Our nervous system is a complex structure that collects information and coordinates activities from all over our bodies. The nervous system contains two main parts: the central nervous system and the peripheral nervous system.

Central Nervous System

There are two structures in the central nervous system:

Brain. This consists of milliards of linked neurons or nerve cells in the skull and serves as the organizing core for almost all of our body functions. It is our intellect's seat.

Spinal cord. This is a packed network of nerve fibers that connects most of our body with our brain.

Peripheral Nervous System

All nerves outside our brain and the spinal cord are made up of the peripheral nervous system. It can be classified into two separate systems:

The somatic nervous system (voluntary). This system helps our bodies and minds to interact with. The somatic system helps our brain and backbone to send signals to our muscles and send information from your body back to your brain and backbone.

The autonomic nervous system (involuntary). This is a system that scans the inner organs and rhythms like the heart, lungs, and digestive system. These are essentially the things that our body does without us having to think about them intentionally. For starters, we can breathe without ever worrying about breathing.

Reading Danger Cues

Our autonomous nervous system (the unconscious system that helps regulate breathing, heart rhythm, digestion, and salivation) is complex and active. Our autonomous nervous system also allows us to search, perceive and respond to threat, in addition to performing such important functions in our bodies as helping us breathe, beating our heart, and helping us digest food.

Sympathetic nervous system. This system encourages our bodies to respond by mobilizing us to step in dangerous situations. Some consider this mechanism a motivation for our responses to battle or flight too dangerous situations in our world. It also stimulates our current medications, known as an adrenaline rush, to pump epinephrine into the bloodstream. When we look upon a snake, our sympathetic nervous system will sense the threat and urge our body to respond, perhaps by rapid adrenaline rush to move away from the snake immediately.

Parasympathetic nervous system. This helps calm our muscles, conserves energy when things start to go sluggish, controls our metabolism, and reduces blood pressure. Some people call it the "rest and digest" system. When we start reading, the cue is not threatening. With the aid of our parasympathetic nervous system, our body starts to relax.

The Vagus Nerve

Dr. Stephen Porges, Ph.D. Dr. Porges is an excellent university lecturer, physicist, and proponent of what is referred to as the Polyvagal principle. There is a nerve, in particular. The vagus nerve is the 10th cranial nerve, a nerve that begins at the medulla oblongata, very long and moving. This part of the brain, the medulla oblongata, is situated in the lower part of the brain. It is situated just above the spinal cord of the brain. This vagus nerve has two sides, the dorsal (back) and the ventral (front). The two sides of the vagus nerve run down across our bodies to the widest range of the nerves in the human body.

Scanning our Environment

From the moment we are born, we search our world intuitively for protection and danger. We are linked and trained for detection, analysis, and reaction to our environment in order to help us survive. A baby reacts with the parent or caregiver to secure sensations of closeness. Additionally, an infant may respond to questions seen as frightening or threatening, such as an intruder, an awful noise, or a lack of response from its caregiver. We search for health signals and risk our entire lives.

Perception

Dr. Porges describes the process of polyvagal philosophy, in which our neural circuits interpret signs of risk in our surroundings as neuroception. Through this neuropeptide mechanism, we view the world in a way that involuntarily tests conditions and decides whether they are healthy or risky.

When part of our adaptive nervous system, this process takes place without even knowing it. Just as we can breathe without consciously suggesting to take a breath, our surroundings can be checked for signs without asking ourselves to do so. During this perception process, the vagus nerve is of particular interest.

Both sides of our vagus nerve can be stimulated during neuroception. Each side has been found to react differently when we scan and process information from our environment and social interactions.

The ventral (front) portion of the vagus nerve reacts to environmental safety and experiences. This encourages physical safety emotions and is socially safely connected to others in our social environment.

The dorsal (back) portion of the vagus nerve is dangerous. It pulls us out of sight, out of consciousness, into a state of self-protection. In times of extreme hazard, we can shut down and feel frozen, an indication that our dorsal vagal nerve is over.

Three Developmental Stages of Response

Porges describes three evolutionary stages in the development of our autonomous nervous system as part of his polyvagal theory. Instead of simply suggesting that our sympathetic and parasympathetic nervous system is regulated, Porges explains the continuum of responses in our autonomy.

Immobilization. This is defined as the oldest road to immobilization. As you may know, the dorsal side of the vagus nerve responds to extreme danger signals, forcing us to sit still. So we will respond to our fear by being frozen, consumed, and shut down. Almost as if we are overdriving in our parasympathetic nervous system, our answer actually means that we stop rather than merely slow down.

Mobilization. In this response, we are drawn into our pleasant nervous system, which, as you may recall, is the mechanism that enables us to mobilize against a dangerous signal. We take action in our adrenaline rush to avoid danger or fight against our challenge. Polyvagal theory suggests that the evolutionary system was about to follow this direction.

Social engagement. It is centered on our main (front) side of the vagus nerve, the latest addition to our list of responses. Recalling that this section of the vagus nerve leads to sensations of protection and attachment, social involvement helps us, by this ventral vagus route, to feel grounded. We can feel safe, relaxed, connected, and dedicated to this room.

The Response Hierarchy in Daily Life

When we live in the world, we naturally face moments where we feel safe and others, or when we will feel uncomfortable or dangerous. Polyvagal theory suggests that we have a dynamic space and that we can travel within the hierarchy of responses in and out of these various places. Social engagement in taking up a safe loved one may occur to us and, within the same day, organize us when we face danger like a rabid dog, theft, or an intense confrontation with a coworker. Sometimes we can read and respond to danger and deal with the situation in such a way that we feel trapped and unable to get out of it. The body responds in those moments to heightened sensations of risk and anxiety, going into a more predominant region of immobilization. Our dorsal vagus nerve is damaged and traps us down to a freezing, numbness and dissociation, as some scientists believe. In those moments, the dangers can be overwhelming, and we see no viable way out. Moments of sexual or physical violence could be an example of this.

Impact of Trauma

If somebody has undergone traumas, particularly when they have been left immobilized, their ability to search their environment for unsafe signs may become distorted. Of course, our body's goal is to help us never suffer like that again, so it's going to do whatever it has to do to protect us. When our surveillance system goes into an over-drive, working hard to protect us, it also can read many of the signals that are harmful in our world–even the signs that are friendly or benign to others. Our social commitment allows us to interact with others more fluently and to feel connected and secure. If our body finds a clue to an interaction that indicates that we may not be safe, it begins to react. That sign will take many people to the point of agitation, which can spring into action and try to neutralize the threat or get away from the threat. For those with PTSD, the threat signal may immediately shift them from social participation to immobilization. By combining numerous interpersonal signals that are dangerous, such as a slight change in facial expression, a specific tone of voice, or certain types of body posture, they can return to a place that is familiar to them to prepare and protect themselves. The body can not register an answer to mobilization as an option. This can be very frustrating for trauma survivors, not understanding whether their experiences with other people and the world affect the structure of reaction.

Connection and Polyvagal Theory

While the vagus nerve is known to be widely distributed and connected to a variety of sections of the body, it is important to note that the system may influence cranial nerves, which control social participation through facial expression and vocalization. As individuals who are related, we can understand how often scans for dangerous signals can occur in our encounters with others that are meaningful, or relevant, in our lives. We strive innately for feelings of safety, trust, and comfort in relationships with others, and we quickly find out what indicates when we are not safe. When people become more comfortable in and for each other, creating strong relationships, expressing vulnerability, and feeling trust can be easier.

Do Not Go Yet; One Last Thing To Do

If you enjoyed this book or found it useful, I'd be very grateful if you'd post a short review on Amazon. Your support really does make a difference, and I read all the reviews personally so I can get your feedback and make this book even better.

Thanks again for your support!